ANIMALS OF THE DARK

ANIMALS
OF THE DARK

Clive Roots

DAVID & CHARLES

NEWTON ABBOT   LONDON   VANCOUVER

0 7153 5965 7

Set in 11 on 13pt Baskerville and printed in
Great Britain by Biddles Limited Guildford for
David & Charles (Holdings) Limited
South Devon House   Newton Abbot   Devon

Published in Canada by
Douglas David & Charles Limited
3645 McKechnie Drive West Vancouver BC

Contents

Introduction

My most vivid introduction to the world of darkness occurred a decade ago in the vast forests north of the Amazon in Guyana. After a wearying day negotiating snags on the Kukui River, I lay awake listening to the night sounds, a veritable hubbub after the stillness of the tropical forest by day. Bats were plentiful beneath our tarpaulin strung between two trees to form a simple roof over the hammocks of my Amerindian companions and I. Roaring in the distance, neighbouring troops of howler monkeys defied each other to enter their territories, and above the steady din of the frogs and cicadas a lonely potoo or giant nightjar repeated its monotonous 'poor me one all alone' call. But only the bats occupied my thoughts. Vampire bats to be sure; possibly even rabid ones whose infected bite meant almost certain death.

Collecting rare animals for zoos had led me to this seldom-visited region, and my mosquito net, an essential item in this area, had been left at base camp, a day's journey down river. Consequently I spent an uncomfortable night wrapped cocoon-like in a blanket as protection from the vampires. A slit at eye level afforded occasional glimpses of the bats, some large enough to be fruit-eating species, others comparable in size to the small blood-letting vampires, although probably just inoffensive insect-eaters. It was as if we had built our simple rain shelter in the very centre of the bats' freeway, and scarcely a

Introduction

minute passed without the slight sounds of them avoiding our suspended bodies increasing my apprehension. Possibly I was over cautious, but British Guiana, as it was then known, had a history of vampire bat activities, plus loss of life from the paralytic rabies they transmit. Under these circumstances even the most impartial traveller could not help feeling uneasy, and in this state of mind the myths and legends of the night take on a new meaning, becoming almost realistic.

Bats may be the most unearthly creatures of the night, but they form only a small percentage of the nocturnal animals whose daily survival routine commences after dark. Man seldom appreciates this; in his world of artificial light, he continues his activities in city flat or organised social surroundings, and after a certain hour considers bats to be unusual in an otherwise sleeping world. However, the setting sun heralds the appearance of a large population of nocturnal animals. By reversing what man considers the norm, and being active at night, these animals are able to make use of niches and habitats occupied by others hours, or even minutes, earlier, without becoming competitors. This is the original night shift, its workers making use of the facilities enjoyed by the diurnal, or daylight, animals, with the exception of direct sunlight, of which most nocturnal animals have little need.

Just as diurnal animals are not necessarily active all day, so nocturnal animals may not be active all night. In general they regulate their waking by the setting of the sun, and their retiring by the sunrise, but much variation occurs according to the species. The twilight hours witness the activity of a specific fauna, known as crepuscular animals, which are not seen during the darker hours. Few of the truly nocturnal animals which rely on sight are abroad during the early hours of morning when the light intensity is at its lowest, as even animals with the most highly developed night vision cannot see in total darkness. Many large animals that are usually active at night also appear during the day for short periods;

this is determined by absence of sunlight, lack of hunting success at night or by the season of the year. Others, particularly smaller species, may divide each twenty-four-hour period into short periods of rest and activity, having a shorter rhythm of activity than, for example, the strictly nocturnal insectivorous bats or strictly diurnal cheetah.

The true fossorial creatures and the troglodytes, or cave-dwellers, have no choice in the matter of preferred light intensity as theirs is a world of permanent darkness. Although they have no use for even the specialised vision of the nocturnal animals, evolution has not by-passed them and adaptations of other sensory organs ensure their survival. Sensory cells along the sides of blind cave fish, for example, record vibrations and changes of pressure, helping them to locate food, and to avoid obstacles and, most important, predators.

Adaptations ensuring survival with minimum light, and in many instances with no light at all, have few equals in the animal kingdom, and at least one of the senses—sight, hearing, smell, taste or touch—is highly developed. There are also minute sense organs in the skin which respond to changes in temperature and pressure, these belonging to the general sensory system. All the sense organs have a similar basic plan, with receptor cells receiving the stimulus and transmitting nerve impulses to the brain, where they are interpreted and a decision made as to the appropriate action to be taken. The sense organs, therefore, are not concerned with the interpretation of a stimulus, only its reception.

There are even adaptations which assist nocturnal animals to survive the daylight hours without harm. The large sensitive eyes of the frogs, for instance, enable them to use all available light, however dark it may appear to human eyes, but their elliptical pupils close to a slit during daylight to protect the delicate retina from bright light. There are definite advantages for the amphibians in being active at night. The lack of dehydrating sunshine allows them free movement after passing the daylight hours in a pool or the moist micro-climate beneath a large leaf. They are safer from attack by reptiles, which are mainly diurnal; and there is an abundance of insect life available. The frogs and many other nocturnal creatures have evolved structural alterations which allow them to burrow into moist places to escape from the sun, and additional modifications enable some to spend all their lives underground.

The echolocation mechanism used by bats to locate their food and avoid obstacles is a completely different form of adaptation. They 'see' in darkness by means of highly specialised hearing which picks up their own high frequency orientation sounds, made at the rate of from five to two hundred per second. The returned echoes are transmitted with amazing speed to the bat's tiny brain, where they are analysed and a decision reached. No insect in flight is beyond the catching powers of an insectivorous bat, which can even discriminate between echoes returned by an obstacle or an insect. When the sun rises, the bats will already have found a secure resting place for the day, often in the cool darkness of caves. In parts of northern South America and on the island of Trinidad lives a remarkable bird with similar habits to those of the bat. Known as the oil bird, or guacharo, it is probably the most highly specialised member of the bird kingdom, using sonar within its cave, which it leaves at dusk to search for oil-palm nuts, its sole diet.

The bat's echolocation powers, the well-developed hearing of the owl, and the highly evolved sensory cells of certain snakes, which can detect temperature changes of a fraction of one degree, all attest to the incredible ability of animals to adapt to seemingly insuperable circumstances. The world of darkness is truly a different world. It is fascinating, mysterious and even frightening to man. This is an introduction to the many interesting creatures of that world, and their way of life. Nocturnal animals, together with those living in the permanent darkness of the soil and caves, amount to over 50 per cent of all the living vertebrates. It is strange that library shelves seldom reveal literature about them, yet are overloaded with books on those animals which, like man, rise with the sun.

CHAPTER 1

Darkness and Life

In the following chapters I have attempted to answer two basic questions. Firstly, how do the nocturnal and subterranean creatures cope in darkness? Secondly, what do the nocturnal species do during the day? How they cope is mainly a story of combined structural modifications and development of the senses, often to the point where man's receptors seem feeble by comparison. For most species, concealment during the day, whether from the sun or from predators, is the natural complement of nocturnal activity. So well do they hide when man is abroad that perhaps his only glimpse of the strictly nocturnal animal was the hedgehog he once ran over in Devon or the quick flash of a wild carnivore's eyes as the animal paused briefly to face his car headlights before disappearing into the darkness.

The animal world of darkness is one of predator and prey, the night-time equivalent of the more familiar diurnal world of the hovering hawk or fleet-footed cheetah. Whatever their patterns of activity, however, animals cannot exist in the highly competitive world of nature unless their senses are highly attuned. To survive, the predators must kill regularly and the prey must escape regularly; and the degree of acuteness of a nocturnal animal's senses depends upon its reliance on those senses. Whatever it is doing, whether feeding, migrating, seeking a mate or even sleeping, it must be constantly

13

alert. Seldom is it possible to approach to within touching distance of a sleeping wild animal; its position typical of the species, its sense of smell or hearing, or even its awareness of ground vibrations produced by footfalls are sufficient to arouse it. In addition to their five major senses, most animals rely upon automatic responses, such as reflexes and instincts—the first being the involuntary responses to nerve stimulation, the latter being the natural and largely hereditary aptitude of species to respond to environmental stimuli.

THE SENSES

Vision is considered the most important sense in man because his highly developed brain, and ability to co-ordinate eyes and limbs, enable him to make the best use of sight. Three sensations are included in human vision—light intensity, colour and form. In most nocturnal animals colour vision is impossible; the retinas of their eyes are poorly endowed with cones, which perceive and discriminate colour at very high levels of light intensity, but are well supplied with rods, which are effective at low light levels. Multiplying in the rods during darkness, the retina's light-sensitive pigment (rhodopsin, or visual purple—manufactured by combining vitamin A with protein in the presence of darkness) absorbs most visible colours, converting their light into energy. Rhodopsin does not have equal sensitivity to all colours, however; it is most responsive to green, falling off towards blue and red at the ends of the spectrum. As nocturnal animals cannot perceive red or very much blue, these colours have been used in reversed activity exhibits or 'nocturnal houses' in zoos.

Sight

Rods and cones are nerve cells, their particular formation depending upon the species and its habits. Physical and chemical changes occur in these cells when light falls on them, sending impulses to the brain. To nocturnal animals, both darkness and daylight are grey worlds, as the picture transmitted to the brain by the rods is in black and white only. Although rods are far more sensitive than cones, their picture is never as sharp because many rods send their message to the brain along a single nerve, whereas each of the more numerous, but less sensitive, cones has its own nerve fibre going to the brain. Briefly, cones provide acuteness of sight, rods sensitivity to light intensity. Some nocturnal animals have a few cones in their retinas, others none at all; examples of the latter being geckos and bats. The diurnal birds have very few rods and are virtually helpless if disturbed after dark, floundering about with little or no chance of finding a secure roosting place. Most mammals, whether diurnal or nocturnal, are colour blind, which no doubt reflects their nocturnal origins during the age of reptiles. The higher primates, with the exception of the nocturnal howler and the douroucouli, are not colour blind.

To have survival value at night, eyes must be capable of receiving sufficient light, and the truly nocturnal species which rely mainly on sight have enormous eyes. The insect-hunting spectral tarsier, for example, has the largest eyes in relation to its size of all the mammals. Nocturnal vision is necessarily myopic or short range, most hunters being able to rely on this sense only when they are within striking distance. The long tubular eyes of the owls result in their being short-sighted and for this reason they always hunt close to the ground.

The animals we are concerned with—the vertebrates—all have eyes of basically the same structure, although in many instances they may have degenerated or been lost completely through lack of use. We usually consider the sense of sight to be the formation of distinct images on the retina, not just the ability to distinguish changes of light intensity; yet some nocturnal animals, the blind fossorial species and the troglo-

Racoons on the
bank of the
Weeki Watchie
river in Florida.
Searching for
crayfish in shallow
water with their
sensitive paws,
they are rarely
seen during
daylight

Darkness and Life

dytes distinguish between light and dark by means of the light-sensitive organs in their skin, which trigger off impulses to their muscles, causing them to seek darker places. Most nocturnal animals can see fairly well in daylight if they are allowed to get accustomed to light slowly, but protection of the sensitive eye is of major importance to these animals during the day. By means of pupil contraction the lens is shielded from glare damage by the pigmented light-resistant barrier known as the iris. In contraction, the shape of the pupil varies; there are two basic types: the pinpoint and the slit, the latter being either vertical or horizontal. Except for the subterranean and troglodytic creatures whose eyelids have fused together, and the snakes which have only an immovable transparent window covering the eye, most terrestrial vertebrates have, perhaps surprisingly, three eyelids. The third, an almost transparent membrane, works at right angles to the top and bottom lids from the inner corner, and is known as the nictitating membrane. It is present in

A nocturnal insect-hunter, the slender loris' large eyes allow sufficient light to enter even on the darkest nights

lizards, crocodiles, turtles and amphibians, allowing good vision while still protecting the eye under water or below ground. Man's third eyelid is now only present in the form of a vestigial piece of pink skin in the corner of the eye. Most carnivores still have a third eyelid, and the nocturnal birds use theirs for protection against bright sunlight, individually if necessary. In the prey species, it is important as a protection against predators as it helps to conceal the iris, which is usually bright yellow and very prominent. Where the actual eyelids and their closing are concerned, in most birds the bottom lid moves upwards when the eye is closed, whereas in mammals the reverse is the case. A bony surround protects the eye in most vertebrates, but in some amphibians the eye socket is bottomless so that the eyes protrude below the roof of the mouth and are depressed when prey is swallowed.

Nocturnal habits place restrictions on the use of sight in general, however well developed it may be. At least where most mammals are concerned, more reliance must be placed upon hearing and smell. Lack of good eyesight, when other senses have been improved upon, is no hindrance for it must be appreciated that the *means* of sense reception is not important. By whatever means it reaches the brain, it is the *message* which is deciphered to build a mental image. Distinctive voices are characteristic of most nocturnal animals and it naturally follows that the sense of hearing must be well developed in these species. Communication between the gregarious nocturnal animals is more reliant upon sound than in the diurnal species, where sight may play a larger part. Crocodilians roar at night, and the hooting and screeching of owls and the night calling of migratory waterfowl are good examples of nocturnal communication. At night in the forests there is additional need for power of voice; the calls of the howler monkeys are so great that their nocturnal roaring can be heard several miles away. The booming of the douroucouli —the only other nocturnal monkey as opposed to primate—

is out of all proportion to its size. Solitary animals as a general rule are less vocal, even if they possess the facilities, as they seldom need to communicate with their own species except at breeding time or when outlining their territory.

The sounds produced by night animals as a means of expression vary according to the species concerned. They can be vocal in origin, as in the examples already quoted; they may be nasal, as in some salamanders; dermal, as in the beak clapping of owls, tusk gnashing of peccaries and rattles of rattlesnakes; or they may be produced by physical actions, such as the hoof drumming of deer and the stomping of porcupines.

Hearing

The ear reaches its highest degree of development in the

Like all snakes, the boa constrictor lacks ears and cannot hear in the human sense. Very sensitive to vibrations, however, it interprets footfalls as potential enemy or prey

mammals and in most species is accentuated by the possession of an outer ear, or pinna. The nocturnal species depending more on their hearing than any other sense, eg the bushbaby, fennec fox and the insectivorous bats, have a greatly enlarged pinna to collect and funnel sound waves into the auditory meatus, or channel of the ear. This meatus is absent in all other vertebrates except birds; in fish, amphibians and reptiles other than the snakes, the eardrum lies on the surface in a shallow depression. Snakes cannot hear, in the anthropomorphic sense, as they have no ears, but they are more sensitive to vibrations than man; they detect these from the ground, particularly through their lower jaws, and interpret them as possible predator or prey according to the intensity of vibration. In the fossorial or subterranean mammals, the outer ear is reduced or completely absent to aid movement through the soil; and in fossorial animals generally, degeneration of the ear cavity has occurred.

The reception of stimuli occurs in the inner ear, which is therefore the most important part of the hearing sense. It has a similar basic plan throughout the vertebrates, but is more complicated in the higher forms, which consequently have a better sense of hearing. Fish have an inner ear only, but the amphibians, reptiles, birds and mammals possess a middle ear which improves the ear's sensitivity. In the mammals and birds an additional structure—the external ear—increases the ear's reception of sound to produce the greatest development of the hearing sense, which is further increased in most mammals by their outer ears.

Sound vibrations, whether through the air, water or ground, are picked up by various animals depending upon their frequency. Some, like man, are adapted to receive low frequency vibrations and cannot hear the high frequency sounds emitted by the bats. No animal is capable of hearing all sounds, but each has a range of hearing varying from a few vibrations per second to many thousand. Individual species may be more sensitive to sounds within a particular area of their range. It is known that cats, for example, are more sensitive to sounds within the higher reaches of their hearing, which extends to about 45,000 cycles per second, and as a result can hear the high-pitched squeaking of small rodents. Man's range commences at 20cps, below which hearing is replaced by the sense of touch, and extends to 20,000cps, beyond which he is deaf. Bats produce sounds inaudible to the human ear without artificial aids and are themselves sensitive to frequencies up to almost 100,000cps. Dogs also have a sense of hearing far beyond man's range and shepherds have long used high-pitched whistles to control their canine assistants.

Smell and taste
Like the sense of hearing, the use of smell is of major importance to many nocturnal animals. The sense of smell is a chemical one, in that particles carried in the air or in water can be detected by the scent or olfactory organs and then related to predator, prey, possible mates or intruders of the same species. It is closely connected with taste, the other chemical sense, particularly in the amphibians. Its advantages lie in providing information, in advance and from a distance, on factors likely to affect the animal or its environment. Coupled with the stimulation of the olfactory organs, there is therefore always a physical response, either towards or away from the source of smell. The resulting increase or decrease in the strength of the smell enables its direction and distance to be ascertained. Although many mammals rely mainly on their sense of smell, the olfactory organs are developed to their finest degree in the lower vertebrates. With a few exceptions, such as the petrels, albatrosses and kiwis, birds are not considered to have a sense of smell.

In the mammals the olfactory nerve endings are situated in the upper part of the nose, the lower part being mainly concerned with filtering and warming the air they breathe.

The highly sensitive, elongated snout and tactile whiskers of the Malayan gymnure or moonrat, aid its night-time search for worms, fish and frogs. Not a true rat, but an insectivore, it is related to the hedgehogs and shrews

Consequently, sniffing as opposed to breathing normally draws more air and therefore more scent particles to the back of the nose, which is why animals hold their heads high and sniff to locate the source of a smell. An additional sense occurs in snakes, lizards and some mammals, which is described as 'smelling the food in the mouth' by means of Jacobsen's organs, whose function is explained on pages 50–1. This sense is of particular importance to the nocturnal and burrowing reptiles.

Taste is also a chemical sense closely allied to smell, the taste buds consisting of groups of sensory cells bearing short taste 'hairs'. These are situated in the buccal cavity, on the tongue and even on the lips of some fish. Taste buds are stimulated by higher concentrations of chemical substances than are the olfactory organs, and result in secretions of the salivary and mucous glands, the act of swallowing and responses in the alimentary canal. They have been described as the second stage in the process of chemical discrimination, second to the sense of smell. In the subterranean animals, where sight is lacking or can be of little importance, and in those animals most active during the darkest period of the night, the senses of smell and taste take on a major role.

Touch

One major sense remains, the sense of touch. It is well developed in the fossorial animals, the burrowers and those amphibians which hide by day in or beneath anything offering escape from the sun. They are referred to as being positively thigmotactic because of their need to be in close contact with their surroundings. In many species, such as the moles, moonrats, hedgehogs and solenodons, highly tactile snouts, strong enough to probe for invertebrates but with a delicate sense of touch, are used to locate food. The racoons have very sensitive fingers with which they locate crayfish and other choice items at night beneath stones in shallow streams and pools. In some aquatic amphibians, such as the pipa and clawed toads, delicate, elongated fingers aid them in locating and seizing their food. Cats' whiskers, or vibrissae, have such sensitive endings that prey animals touching them are immediately seized.

Hiding by day is of the utmost importance to most nocturnal animals; surprisingly, it is not only the thin-skinned moist amphibians which are vulnerable to sunlight exposure. Even the crocodiles with their 'armour-plate' hides cannot lie in the blazing sun for long, and experiments have proved that exposure for less than an hour can be fatal to them. Under normal circumstances they frequently enter the water to lower their body temperature and often have a protective layer of mud over their bodies. Keeping their jaws wide open and panting also expels heat and reduces their temperature. Even snakes which love to bask cannot withstand exposure to direct sunlight for very long; experiments have proved that grass snakes and allied species may succumb from half an hour's exposure to sunlight on a hot day.

Although some nocturnal animals bask in the sun, others find daylight quite unbearable and are obviously uncomfortable if they cannot seek shelter, eventually protecting their eyes by turning away from the source of light or by curling up with their heads covered. When exposed to daylight, the kinkajou's first thought is to seek cover; and at night it reacts to bright light by blinking and peering through half-closed lids, apparently unable to protect the lens by pupil contraction. The opossums, lorises and pottos are similarly very sensitive to bright light.

CHAPTER 2

Troglodytes and Troglophils

We depend so much on our powers of vision that it is difficult to imagine sightless animals, some blind from birth, living a 'normal' life. In many forms of vertebrates, however, vision is not always the most important sense. In the fossorial animals and the cave-dwellers sight has so degenerated that it is absent in many species. Despite lacking sight, as we understand it, many are sensitive to light even though the reception of a definite image is impossible. Even among the higher vertebrates—the mammals—there are species, living above ground, where the senses of smell, touch and hearing play a greater part in their lives than sight. The weasels are excellent examples; they have such well developed olfactory organs, compared with their powers of vision, that they could almost certainly survive if deprived of their eyesight. Lower down the family tree, the barbels of the catfish and the long sensitive fingers of some aquatic amphibians are of more use to them than vision in locating food. The snakes, and particularly the pit vipers, have such well developed sensory adaptations that they can locate their prey in complete darkness. It is not so unusual, therefore, that the vertebrates living in the permanent darkness of caves and the fossorial reptiles and mammals should manage without sight.

Troglodytes and Troglophils

CAVE FISH

The blind cave-dwelling vertebrates we are concerned with are fish and salamanders; compared with the eyed species, the number of sightless forms is relatively small. The troglodytes are those found only in cave systems, and therefore have degenerate eyes and little or no pigment. As light never enters their world, most do not even have light-sensitive cells in their skin. Troglophils live deep in caves too, but also occur at their entrance and outside; depending upon their habitat, they may have eyes and pigment, although in reduced form. Possibly the most unusual aspect of the cave fish is not so much that they are sightless, but the fact that their young are born with eyes which degenerate as the fish grow. This is all the more unusual when we consider the results of experiments conducted with goldfish kept in complete darkness. The young born to these fish were sighted, but gradually lost their sight and after a few years all the young were born without eyes. When the goldfish were returned to the world of light, they did not regain their sight, but also produced blind offspring. This in itself is highly unusual, as acquired characteristics of any kind are not generally passed on to the young. Including the deep sea forms, it has been estimated that there are about eighty species of blind fish, but only a few of these are freshwater cave species. Most of the blind fish have been discovered in fairly recent times and new species are still coming to light as cave systems are explored.

The blind fish are thought to have been pre-adapted to life in the caves before they left the outside world. No doubt they already had well developed sensory cells and must have possessed the ability to make the necessary endocrine adaptations. Darkness and the constant temperature of cave water probably had physiological attractions to them, as they prefer and seek warm water. They are said to be positively thermotrophic— able to detect very slight variations in water temperature.

Troglodytes and Troglophils

The best-known blind fish is the well-studied Mexican blind or cave characin, discovered in 1936. Most troglodytes have an ancestral relative above ground, which usually differs only in possessing eyes and pigment. Searchers in the Mexican caves have located a whole range of cave characins, from pigmented, eyed forms near the cave mouth to completely blind, pink forms in the darkest depths where light never penetrates. The Mexican blind characin is the only blind cave fish which reproduces regularly in captivity and has therefore been the object of extensive scientific investigation. Surprisingly, it was found that the sightless forms are sensitive to light, no doubt as a result of their vestigial optic nerves. They are thought to have first entered the caves many years ago in search of the volcanically warmed water, and for safety from predators.

Mammoth Cave in Kentucky revealed its blind cave fish in 1842, and these are believed to be descendants of fish which entered the vast limestone caves of the central USA many millions of years ago, possibly in the Tertiary Period. Four blind species have been discovered in this region, all belonging to the genus *Amblyopsis*. One of these, the spring cave fish, is a troglophil, found in springs outside the caves as well as in the blind form. Experiments conducted with the swampfish— the eyed and pigmented troglophil ancestor of the Kentucky blind fish species—have shown that evolution underground resulted in modification of the ductless glands, as these fish developed pituitary imbalance, affecting their skeletal development, when kept in the dark.

Despite lacking external ears, fish are capable of hearing, as their extremely sensitive inner ears, which receive sounds passing through the water, are situated close to their brain. Water conducts sound better than air since it is incompressible; as fishes' bodies consist of 90 per cent water, sound waves pass directly into the fish and its inner ear without needing amplifiers. Vibrations play such an important part in the lives of fish generally that the females of some species will not lay

27

until they sense the vibrations of a male, which need not be in actual physical contact. The reason they are so sensitive to vibrations is because they possess lateral line organs, which take the form of a tube—the lateral line canal—running along each side of the body, between head and tail, and opening to the water through numerous pores. This tube contains receptors, known as neuromasts, which have hair-like projections extremely sensitive to low frequency vibrations, informing the fish of even the slightest movement in the water near them. Movement of large objects such as possible predators causes them to scatter, whereas lesser vibrations are attributed to possible food items. To fish lacking sight, it is obvious how important these lateral line organs are to their survival.

The *Amblyopsis* cave fish of the Central American limestone caves have exposed neuromasts on their heads and bodies, which increases their sensitivity to underwater vibrations. They also have lengthened fins allowing them to swim 'silently', thereby reducing their own vibrations which could interfere with the reception of others. They carry and incubate their eggs and small fry in their gill chambers, ensuring a high hatching and survival rate of the young.

In the caves of Cuba two species of blind fish have been discovered, but have yet to be given generally accepted common names. As an additional aid, these fish have sensory cavities on their heads and, although their lenses begin to degenerate before birth, they are actually born with eyes which then continue to shrink, the eye muscles degenerating throughout their lives. It is thought that their ancestors lived in coral reefs, as the members of the Brotulid family to which they belong are typically marine, and became land-locked as the coral rose above sea-level and formed an island.

Another blind and unpigmented fish which is still marine, occurring in the intertidal caves of Baja, California, is the blind goby; it lives in burrows in company with an equally blind and colourless shrimp. Unlike their relatives from the world of light, these gobies lack spines and have small fleshy projections around their mouths which serve as organs of touch. Gobies are commensals, relying entirely upon the shrimps for board and lodging without causing them harm. The shrimps dig a burrow and live at its entrance, allowing the gobies to live inside it. The shrimps' continual movement wafts food and oxygen into the burrow on the water currents and when the shrimps die the gobies are helpless.

CAVE-DWELLING SALAMANDERS

As caves are usually of fairly constant temperature and humidity at all times, they provide an ideal habitat for salamanders. These tailed amphibians do not require very high temperatures but need high humidity, a condition which exists permanently in many large caves. Their life cycle is lengthened in caves as the lowered temperature results in their eggs taking three months to hatch, instead of the usual two as is the case above ground in warmer temperatures. With the exception of the olm of the Proteidae family, the cave-dwelling salamanders belong to the Plethodontidae family the members of which are completely lungless. They have no nostrils but breathe through their skins and the lining of the mouth, carbon dioxide being similarly voided through the skin, which must be moist at all times. They are also peculiar, or at least different to other animals, in that the lower jaw is immovable, despite being hinged in the normal way, and the mouth is opened by lifting the upper jaw and head.

A highly developed sense of smell is important to animals living in the permanent darkness of caves. Even their larvae, which may live in relatively clear water outside the caves, normally feed at night when sight is of little use and a sense of smell important. Experiments which involved blocking the nostrils of tailed amphibian larvae left them completely insensitive to the presence of food. A good sense of smell is also essential at breeding time as the salamanders are virtually voiceless. The male lungless salamanders have a gland on their chins, from which a mucous secretion is discharged; this appears to have intraspecific importance, attracting individuals of the same species during courtship. Salamanders smell with the cells inside their nostrils and with specialised cells in the nasal chamber, collectively known as Jacobsen's organ. Obviously the olfactory organs, their standard sense of smell, must be highly developed. The cells of Jacobsen's organ test the food when it is in the mouth, providing information to the brain on its palatability.

In some amphibians, particularly the cave-dwellers, the pineal organ, or vestigial central eye, is believed to function as an endocrine gland, since it is no longer needed for light reception. In all amphibians, basically nocturnal anyway, the pineal eye has ceased to function as a light receptor. Salamanders do not have ears and, as they are virtually incapable of uttering a sound, the lack of a sense of hearing is no great drawback. They are said to be able to hear airborne sounds only if these are below 244 cycles, receiving the vibrations through their bodies. Just as the burrowing and fossorial animals seek maximum tactile sensation at all times, the larvae of some amphibians have such a highly developed thigmotactic instinct—needing to be in close contact with their surroundings in a crevice or cavity, for instance—that this overcomes their dislike of bright light, and they may lie in a narrow cavity fully exposed to light from above. Normally they are negatively phototactic, meaning that they seek darker places when exposed to light.

Several species of blind salamander also occur in the limestone caves of the southern USA. Living in perpetual darkness they lack pigment and their internal organs can clearly be seen through their translucent pink or white skins. Like the blind fish from the same region, it is believed that their ancestors first sought the protection offered by the caves millions

A larva for life, the blind olm lives in caverns in Yugoslavia, where the coldness of the water prevents it completing its metamorphosis

of years ago. The larval stage of the grotto salamander lives in mountain streams, is pigmented, has functional eyes and a high tail fin. When they have completed their metamorphosis and attained adulthood they move into the caves where they slowly lose their pigment and tail fin. Their eyes are not needed either and the eyelids fuse. They remain in these caves for the rest of their lives, and although their breeding habits are unknown it must be assumed that the eggs of newly hatched larvae are washed out in the streams, as the adults never leave the security of their cave fastnesses. When maintained artificially in natural light, young grotto salamanders have retained good eyesight and pigmented bodies.

From a well sunk 200ft deep in Albany, Georgia, the first known specimen of the Georgia salamander was obtained in 1938. A completely blind species, it is only 3in long, with red plume-like gills on a white body. It does not have externally visible eyespots, in contrast to the Texas blind salamander where the position of the (completely sightless) eyes is indicated by small spots beneath the skin. This species is a spindly legged, ghostly amphibian, about 5in long, with a long flattened snout. It retains its gills throughout life, breeding in the larval stage. It is found only in the deep caves of the Balcones escarpment of central Texas, where the larvae live in surface streams and have normal eyesight, but migrate to the caves as they are about to metamorphose.

Within the family of lungless salamanders there is a genus or group known by their scientific name as the Hydromantes salamanders, which have sometimes been called the web-toed salamanders. This group contains the only other known cave-dwellers in the family, but they are pigmented and possess eyes, living mainly near the entrances of caves in dim light. A flattened body and webbed toes are characteristics of these creatures which are very agile yet slow in their movements. A mushroom-shaped tongue, which can be protruded for one third of its body length, aids in the capture of its invertebrate food. Of the six known species, three occur in California, of which only one is a cave-dweller; the others live in Europe, where two species dwell in caves. The North American cave species is the Shasta salamander, named after the limestone caves of Shasta County, California, where it was discovered. The two European cave-dwelling species are the only known members of the lungless salamander family in the Old World. Gorman's salamander has only been known to science since 1952, when it was found in caves in Tuscany and has yet to be found elsewhere; and the Italian salamander is found in high altitude caves in southern France and northern Italy.

Also in Europe, in the permanent darkness of caves in Yugoslavia, lives the olm, a salamander which retains its larval form throughout life. It is possible that the low temperatures of these caverns prevents them from metamorphosing, as it is known that coldness inhibits the release of thyroid hormone which controls metamorphosis. They are blind and lack pigment in their 12in-long bodies, but turn black when exposed to normal light. Contrasting with their colourless bodies are the bright red gills which they retain throughout life.

CHAPTER 3

Amphibians in the Shadows

About 350 million years ago the first daring fish-like animal crawled out of the water and initiated the struggle to colonise the land. It was a long task. Fins had to be transformed into limbs; eyes which had evolved over a long period for underwater sight had to be adapted for life above water, and more reliance had to be placed upon other senses instead of the sensitive lateral line organs which were unsuitable to the new conditions. The most important stage—the development of lungs from an air bladder—must have commenced earlier, and the first amphibian probably functioned like the present-day lung fish which has both lungs and gills and can survive on land for several months. All modern amphibians still pass through a gilled stage, although a few lose their gills before they leave the egg, and despite possessing lungs, most species still depend upon their skins for much of their oxygen supply. Acting as a respiratory surface, the moist skins of frogs, for example, are thought to provide at least half the animal's oxygen. The lungless salamanders have no alternative, however, and must breathe at all times through the lining of their mouths and their very thin skin, which has a rich supply of capillaries almost reaching the surface.

The majority of living amphibians are nocturnal; when active at night they avoid the excessive temperatures of the day, hiding before daybreak under anything offering shelter from the sun. Rotting tree stumps, rock crevices, damp soil and leaf mould are favourite locations for salamanders, the discharge from their naso-labial grooves helping to keep their nostrils clear of soil. The small delicate-skinned treefrogs survive the daylight hours attached to the underside of large leaves, where the humid microclimate protects them even when the sun's rays strike the upper surface. Escape from the sun is so essential to the amphibians that many diverse habits and modifications have evolved. Water storage, migration, and burrowing aids are examples of adaptations which assist the terrestrial frogs and toads to survive unsuitable conditions.

As a preliminary to croaking, the painted reed frog expands its throat sac. Active at night, this moist-skinned frog avoids the dehydrating temperatures of the African day

The axolotl, larval stage of the tiger salamander, often fails to meta-morphose in some regions of the United States. Fortunately it is capable of reproduction in its larval form

Parental care of the eggs and young serves this purpose also, in addition to protecting them from predators. Only three orders now exist within the class Amphibia. The members of the first of these, the caecilians, are fossorial in their habits and are included in chapter 10. Of the second order—caudata, in which a tail persists throughout life—we are interested in the salamanders. Many frogs and toads of the third order—salientia—are nocturnal and provide the bulk of the individuals covered by this chapter.

MEANS OF SURVIVAL

Most amphibians pass through an aquatic larval stage in which their gills absorb oxygen from the water, which must be fresh or in some instances slightly brackish. Salt is taboo to frogs and toads, although a few forms, of which the giant marine toad is the best example, survive in estuaries and coastal pools topped with salt spray from high tides. In most species meta-morphosis occurs, during which the gills are absorbed and lungs begin to function, but some retain their gills throughout

35

life and are able to breed in this adult 'larval' form. Neoteny, as this persistence of the larval form is called, is thought to occur mainly because of insufficient iodine, as is shown by the high metamorphosis failure rate in certain lakes of the western United States which lack this element. Most of the western races of the tiger salamander do not metamorphose, except in areas where the larval stages are vulnerable due to the seasonal drying of the lakes. When breeding time arrives the latter migrate at night to water, but for the rest of the year are seldom seen near it.

The amphibians' metabolic rate depends upon their body temperature and, even when they are most active, is far lower than that of the homoiotherms, or warm-blooded animals. During aestivation and hibernation it drops even further. They are able to maintain their body heat in a warm and humid environment, as there is no evaporation of moisture

Toads generally can survive in drier conditions than frogs. The American toad burrows into soil or beneath rocks to aestivate during dry spells

from their skin, but in dry conditions evaporation and heat loss occur, resulting in lowered metabolism and inactivity. Amphibian life is therefore dependant upon moisture, not necessarily open water but moist places, for all stages of the life cycle. Water is essential for reproduction and the survival of young where most species are concerned, exceptions being the viviparous salamanders and certain frogs and toads which lack a free-living larval stage, their tadpoles undergoing metamorphosis in the egg. The larvae of some species can withstand surprisingly high temperatures, an example being the tadpoles of the leopard frog, which have been found in water measured at 106°F in Yellowstone National Park.

In temperate regions the warm season must allow sufficient time for the development of larvae into frogs and also for food gathering before their period of dormancy. It is thus possible for amphibians to live just inside the Arctic Circle, like the Swedish race of the European frog. Amphibians therefore occur throughout the world, excluding Antarctica, but are most plentiful in the tropics. Many are surprisingly active at quite low temperatures, some salamanders commencing their breeding activity as soon as the ice breaks up in spring. Others have been collected from mountain sides 10,000ft above sea level, and a captive European fire salamander remained active for several days at a temperature of 34°F while being encouraged to hibernate.

The thin-skinned frogs are less able to withstand the effects of drought than are the toads, whose dry, warty skins give them some protection from dehydration, enabling them to live in regions receiving scanty rainfall. There are exceptions to most rules, of course, and the flat-headed frog of the central Australian deserts survives lengthy droughts in its deep burrows, living on water stored in its body since the last rains. Amphibians do not drink, their water supply being taken through the skin, but even during normal arid conditions there is a certain amount of dew at night which they can utilise. However, they cannot exist in the truly waterless deserts, such as Arabia's Rub al Khali, America's Death Valley and the Sahara; but as a result of certain modifications and habits they can survive in regions where a protracted dry season occurs. Secretions of the mucous glands assist in keeping the skin moist, but rather than await death by dessication in times of drought they abandon evaporating pools and marshes and hide in crevices, under stones or wherever moisture is retained. Experiments have proved that toads can live for up to two years without food, provided the conditions are otherwise suitable. The clawed toad, famous for its assistance in pregnancy tests, is a good example of amphibians which aestivate—the summer equivalent of hibernate. Living in South Africa, where many rivers dry up during the summer, they burrow into the river bed and remain there until more favourable conditions return. Escape from frost is equally important for others; an aquatic example in Europe, the fire-bellied toad leaves the water in autumn and burrows below the frost line.

Burrowing to escape the sun, or frost, is a common habit and in several species—such as the Mexican burrowing toad and the thin-skinned spadefoot toads—spades have evolved to aid their digging. These are horny tubercles on the sides of their feet; in some species these are elongated, in others short and rounded, and in a few species wedge-shaped. At night spadefoot toads return to their burrows with ease, indicating that their directional sense is well developed.

Civilisation has aided several species in their search for moist conditions. The Colorado river toads, for example, congregate around wells and cattle troughs in their very dry region, and as a result of this extended habitat are said to be increasing. Possibly species in Australia and the African savannahs benefit in the same way. In the south-eastern United States the greenhouse frog is regularly found in the moist soil beneath plant pots.

A nocturnal aquatic amphibian, the eel-like amphiuma of the southern
United States hides in reeds or crayfish burrows by day

As a further survival aid, the hatching period of some amphibians' eggs may be lengthened if conditions are unsuitable
when they are due to hatch. The eggs of Bibron's frog of
Australia, laid in moist soil, may take ten weeks longer than
normal to incubate in times of drought. The terrestrial marbled salamander, unlike most other species, lays her eggs in a
shallow depression on the forest floor in the autumn, where
they remain until rainfall fills the nest the following spring.
Even the eggs of the completely aquatic amphiuma have
been found in moist soil at the bottom of dried-out pools,
although it is only fair to add that it has not been ascertained
how long they can survive in this condition. The tadpoles of
some species are extremely tenacious of life and even have
their own survival techniques when the sun threatens to
dehydrate them. Spadefoot tadpoles are excellent examples.
They band together when food is short and stir up the mud
at the bottom of their pools in an attempt to find food, and

may even be cannibalistic so that a few may survive rather than all perish. To lengthen the life span of the small amount of water available to them they have also been known to scoop out depressions in which they can gain a few extra hours, or even minutes, to metamorphose and relocate to moister areas.

METHODS OF COMMUNICATION

With one exception, salamanders are voiceless, the only species known to utter a single note by means of its lungs being the Pacific giant salamander of western North America which emits a low rattling sound. Others, such as the lungless arboreal salamander, which may climb 50ft high in the oaks and pines of the Californian coastal ranges, makes a weak squeak when handled by contracting its throat and forcing air through the nostrils. As they lack voices it is not surprising to find that the salamanders have no visible ears, as the sense of hearing in the amphibians is associated with the possession of a voice. The frogs' ability to call during the breeding season to attract mates from considerable distances shows they have a good sense of hearing, except perhaps the Mexican burrowing toad which lacks an external ear. A frog's ears are situated just behind the eye, but what we actually see, as a flat shiny round patch, is the typanum, which is the equivalent of the eardrum in higher animals. It is believed that both the volume of call and hearing sensitivity are related to the urgency with which mates need to be attracted. The spadefoot toads, for example, which have a loud far-carrying call, live in fairly dry country and need to attract mates quickly to make use of the pools formed after a sudden downpour. Allied to this is the very rapid rate of growth of the larvae, which metamorphose within two weeks of hatching. At the other end of the volume scale the large red-legged frog of western North America is a permanent water-dweller and makes a barely discernible guttural sound. Some frogs even have a distinct breeding call, the barking frog of Florida, for example, reduces its many syllabled tree-top call to a single bark when seeking a mate at pool level.

Frogs are apparently sensitive to sounds in the range of 50 cycles to 10,000 cycles per second, compared with the salamanders' sensitivity only to sounds below 250 cycles per second, these low frequency vibrations being picked up in the water by the lateral line organs, which all aquatic salamanders possess. This sense of 'hearing' is also present in the aquatic frog and toad larvae, but exists in no vertebrates higher than the amphibians. Lateral line organs are actually groups of sensory cells which respond to low frequency vibrations, but are less sensitive than those of the fish. They persist in some adult amphibians also, the African clawed toads and South American pipa toads have vibration-sensitive organs along their sides very similar in structure to the lateral line canals of fish. They are able to locate moving food in the water by picking up its vibrations. Frogs are also sensitive to vibrations through the ground, a footfall usually being sufficient to instil silence in a pond full of croakers.

Although not all female frogs and toads are mute, voice is

When a flash flood soaks the south-western deserts of the United States, Couch's spadefoot toads attract mates quickly with their loud calls

Thriving in moist soil and beneath leaves, the spotted salamander of eastern North America breeds early in spring after warm rains

more strongly developed in the males and practically all species attract their mates by vocalising. There is considerable variation in their frequency, duration and pitch, each species having a distinctive call recognised by a potential mate of the same species. The red-spotted toad's call is a high pitched trill, the western spadefoot croaks, Darwin's toad has a clear bell-like call and the pig frog actually snorts like a pig. Tropical and subtropical species call throughout the year as breeding is not seasonal, whereas in temperate regions breeding occurs in spring and early summer, when they congregate in large numbers. Distinctive breeding song is important when several species occupy the same pond, marsh or river and have the same breeding season.

Frogs cannot croak until they are sexually mature, which has been ascertained to be fourteen months for the European frog. In addition to their mating calls they produce a high-pitched squeal of terror, which startles predators and acts as a warning to other amphibians. A third type of call, usually a grunt, is made when a male frog is accidentally grabbed by another male during the breeding season, indicating that it must be released. Their calls are produced by forcing air over the vocal cords, and as they do this with their mouths closed they can vocalise under water and underground. The males of many species have a vocal sac at the base of the mouth, which acts as a resonator and in some species balloons outwards from the throat as it fills with air. The salamanders, which cannot vocalise to attract mates, rely upon their sense of smell, even the lungless species—the skin breathers—having nostrils and well developed olfactory organs. Excretions from glands on the lungless salamanders' chins, derived from the mucous secreting glands, attract females at breeding time.

SENSORY AIDS

A good sense of smell is also important to the burrowing and cave-dwelling amphibians and to the aquatic species which hunt by night or feed in murky water. Frogs and toads also have well developed Jacobsen's organs which 'smell the food in the mouth' in addition to their nostrils. The natterjacks and marine toads have such a keen sense of smell that they often depend more upon this than on vision, both being attracted to refuse dumps by the smell. They are exceptions, however, as frogs and toads in general rely on sight when seeking their prey, determining its palatability with the taste buds which line their mouths, being quick to reject distasteful insects and inorganic objects caught up with their food. Movement stimulates their optical senses, large objects repelling them, small ones attracting them. Even the slight movement of the rib-cage in an otherwise perfectly still mouse is sufficient to draw a toad's attention, but it has no interest in inanimate objects and would starve to death beside a pile of dead flies. Amphibians' skins are light-sensitive, and it was proved many years ago that even frogs, purposely blinded, moved to face the direction from which light was allowed to enter their darkened box. Their eyes, which project from the skull, give them a clear view of approaching danger or food as they can see forwards, backwards, upwards and sideways. Binocular vision is not possible in most salientians at close range, however, and if food is placed under their noses they are forced to back away to bring it into focus. The bullfrogs and treefrogs must have exceptionally good long-distance vision as they are able to locate insects moving towards them and intercept them in flight with their long viscid tongues or by leaping up at them. The gliding frogs of the genus *Rhacophorus* of South East Asia are able to leap 50ft between trees, so they must also have well-developed vision. In contrast, experiments have proved that the terrestrial toads cannot see movement more than 10ft away, and being colour blind they do not recognise red any more than the fighting bull.

There is a great deal of variation in iris coloration amongst the amphibians. Quite often the iris resembles the animals' body colour, an obvious exception being the bright red iris of the green red-eyed treefrog, but expansion of the pupil in the nocturnal species is so great on dark nights that very little iris is visible. During daylight when the pupil contracts to protect the eye from light, in many species the aperture is a horizontal slit, particularly in the true toads, and a vertical one in the spadetoads and certain arboreal frogs. Salamanders generally have larger pupils than the frogs, and smaller, less colourful, irides.

It is interesting to note that some of the long extinct ancestors of the frogs had a third eye, positioned on top of the head. This pineal eye can still be seen in the larval stages of some species, but is no longer functional. It appears as a light receptor, flashing sensations to the brain, but degenerated as

The American bullfrog's eyes project from its skull, giving clear vision forwards, backwards, upwards and sideways. Depressing its eyeballs when swallowing, the bullfrog helps to push food down its throat

most amphibians became nocturnal and had little need for it. The upper part of the frog's eyeball is protected by a thick layer of skin, actually an immovable upper lid, and the lower lid has evolved in some species into a nictitating membrane which can be drawn up to cover and protect the eye while still allowing some degree of vision. Frogs swallow their prey by closing their eyes and depressing the eyeballs which project below the roof of the mouth, forcing the food, with the help of their tongues, into the gullet. All species are carnivorous, existing upon any form of palatable animal life from small insects to young birds and mammals the size of mice. In the completely aquatic forms, such as the pipa toad and clawed frog, the tactile sense is well developed for seeking food beneath stones, in mud or dense weed. The tongueless pipa toad spreads out its long sensitive fingers on either side of the mouth, and attracts prey by vibrating a small muscular flap of skin hanging from the snout. When a suitable morsel comes within reach the nerve endings in the fingers flash a message to the brain and the toad lunges forwards with open mouth.

PROTECTIVE MEASURES

Parental care of the eggs and larvae was briefly mentioned earlier. Many of the techniques evolved, although basically to shield the vulnerable stages from predators, also provide protection from the sun. These are, of course, adaptations mainly of the arboreal species and the moist-earth dwellers, as opposed to those which lay their eggs in water. In some species the male carries the eggs—wrapped around the hind limbs in the case of the midwife toad; in brood pouches by Darwin's frog, or on its back, as is the habit of the poison-arrow frog. The procedure adopted by Darwin's frog is particularly interesting, for as soon as the eggs are fertilised the male stands guard for two weeks, by which time the tadpoles can be seen moving inside their eggs. He then gathers them up, ingesting them into throat sacs which extend down the sides of his body almost to the hind legs. The spawn, nourished through the father's tissue, complete their metamorphosis in his brood pouch, eventually appearing as young frogs.

A species of Hylambates from tropical Africa has a similar technique to Darwin's frog, the female holding the eggs in her mouth until they have completed their metamorphosis, so protecting them through their most critical period. Treefrogs of the genus *Rhacophorus*, commonly known as gliding frogs, produce fluid at the same time as their eggs, and beat this into a frothy nest which hardens on the outside and protects the eggs until the larvae have metamorphosed. The tree-living members of this group must ensure that their nests are placed over open water into which the tiny tadpoles can drop as they hatch. Failure to hit their target would almost certainly result in their dessication. To make matters more difficult in locating water—sometimes little more than puddles—and in positioning their nest directly above it, these frogs are nocturnal and do all their bomb-siting at night. These small frogs have vertical pupils which expand greatly in poor light, enabling them to locate their insect prey and leap from branch to branch way up in the forest canopy where the dense leaf growth efficiently blots out even the brightest moonlight. Probably the most bizarre habit, however, belongs to the male barking frog who guards the eggs laid in crevices. Their incubation period is about four weeks and during dry spells he wets them with urine to prevent their dehydration.

CHAPTER 4

Night-active Reptiles

Although the cool tropical nights do not suit many species, there is still a fairly large population of reptiles, snakes mainly, which prey upon the hordes of nocturnal rodents, amphibians and other night-active animals, plus non-active ones such as roosting birds, sleeping squirrels, monkeys and others. Light controls their activity and they are stimulated by darkness and the relatively low temperatures. At night they must rely upon the air temperature and the radiant heat from sunwarmed rocks and soil. Many bask during the day, but others cannot stand exposure to light, especially direct sunlight. Wagler's pit viper, for example, becomes sluggish when exposed to light, while sunlight is said to have such a tranquillising effect on the kraits that they are unable to bite.

The reptiles control their body temperature by alternately basking and sheltering, thus regulating the absorption of heat. Although they have rather a slender zone of optimal temperature they are, of course, active outside this in their attempts to reach the preferred temperature for activity—hence growth and reproduction; but this varies considerably between species. Some desert-living lizards are active for short periods only when the temperature is about 100°F, but no reptile can survive such exposure for long. It has been shown in captivity, however, that lizards generally have higher preferred temperatures than the snakes, even the sidewinder—a desert rattlesnake—succumbing if exposed to the midday sun for more than a few minutes. In captivity also, continued exposure to temperatures just a few degrees higher than the optimal has proved dangerous to reptiles, and strangely so has long exposure to preferred temperatures from which there was no escape. In short, brief periods within their optimal zone, followed by exposure to lower temperatures seem to be necessary for the snakes, if not all reptiles.

One of the many unique aspects of snakes is that they do not have eyelids, merely an immovable scaly 'window'. Lizards have eyelids and, with the crocodilians, turtles and amphibians, also have an extra eyelid known as the nictitating membrane, which protects the eye while still permitting good vision. The snakes' field of vision is mainly monocular, with a slight overlap, the greatest degree of binocular vision occurring in the bird-eating tree snakes where a second chance at their victim is seldom possible. Experiments to ascertain the power of vision in rattlesnakes proved that they could detect movements from a distance of 15ft, although vibrations were picked up through the ground long before the prey reached this point. Sight plays the most important part in food location in the diurnal snakes and carnivorous lizards, where the movement of their prey stimulates them into action. The vegetarian lizards are obviously not stimulated to feed by movement and must rely upon smell in addition to sight. There is no doubt that some species can perceive colour, as yellow, red and orange flowers are preferred.

EXTRA-SENSORY PERCEPTION

Sight is probably utilised during bright moonlight by the nocturnal species, but on dark nights they must be able to locate their prey by other means. During recognition and courtship sight is important initially but thereafter recognition is more dependant upon the sense of smell, as in most

Banded kraits shortly after hatching. A strictly nocturnal species, they cannot stand exposure to direct sunlight for more than a few minutes

reptiles the sexes are alike in form and colour. Their scent may be increased by a discharge, the nocturnal ringnecked snake, for example, attracts others of its kind with its pungent cloacal discharge.

All snakes and lizards have two senses of smell, the standard one where nerve endings in the nose perform the function in the generally accepted way, and a second form which is connected with their constantly flicking tongues, usually passed through a small opening between the lips. Often thought to be 'stings' in their own right, snakes' tongues have no connection with their killing capacity, their precise function being to collect and transfer scent particles to two pits in the roof of the mouth. These mouth pits, known as Jacobsen's organs, are supplied with nerve endings similar to those in the nose and are also connected to the olfactory nerves. A duct opening towards the front of the palate links the mouth and Jacobsen's organ, the particles being transferred from the tongue to the duct, thence to the sensitive membrane for tasting. Snakes can follow a hot trail by this means, and the vipers—which seize

50

Small sensitive pits in front of its eyes allows the green pit viper to locate prey in complete darkness. Extremely sensitive to infra-red radiation, these pits detect temperature variations of less than one degree

their prey, inject venom and then release their victim—are able to track the dying animal at night.

Jacobsen's organs are used in conjunction with the standard sense of smell, and are well developed in all species except the lizards, such as the diurnal chameleon and anoles, which hunt by sight. The reason for this combined sense of smell is not entirely clear, although it has been suggested that their rate of breathing does not allow air to enter the nostrils fast enough to collect and translate scent particles. Night-hunting snakes include a wide range of non-venomous constrictors and colubrids; mildly venomous colubrids; some of the highly dangerous cobras and vipers, and the majority of the pit vipers.

SNAKES

The pit vipers can be easily identified by the presence of a

30633

sensory pit below and in front of the eye on each side of the head, almost forming a triangle with the eye and nostril. The pit is actually a cavity in the maxilliary bone, covered with a membrane and connected to the outside by a minute opening. The membrane is well supplied with nerves and contains receptor organs which are sensitive to the infra-red radiation emitted by their small mammal prey. It has been shown experimentally that the pit is indeed a sense organ, a sense which no other animal possesses. In the name of science, early experimenters blindfolded rattlesnakes, removed their tongues and plugged their nostrils, and were intrigued to find that they still struck accurately at covered electric light bulbs, confirming that the pits guided the reptiles to the source of heat. So sensitive are their pits that they can detect temperature differences as low as 0.2°C, and can locate sources of infra-red radiation—invisible rays beyond the red end of the spectrum, whose wavelengths are too long to stimulate man's retina and too short to produce the stimulation of heat in his receptors. Widespread through the warmer regions of the world, except in Africa, the pit vipers are ecologically adapted for life in a wide range of habitat. There are terrestrial forest forms, such as the 10ft-long bushmaster; prehensile-tailed arboreal species, the largest of which is the 4ft-long Nicobar Island pit viper; desert-dwelling forms like the sidewinder, and semi-aquatic species like the moccasins. They occur from sea level, like the fer-de-lance, to 16,000ft up in the Himalayas—the highest recorded snake habitat—where the Himalayan pit viper is found. The majority are live bearers, the bushmaster being the only American example which lays eggs, while in the Old World the Malayan moccasin is an egg-layer.

All the pit vipers are highly venomous reptiles which account for many human lives annually, and their appearance does nothing to dispel man's apprehension and hostility towards snakes generally. In common with other snakes they have 'cold' expressionless eyes and lack eyelids, but their sinister appearance is enhanced by their vertical pupils, an enlarged and overhanging scale over the eye, and a heavy diamond-shaped head to house the enlarged venom glands. Their fangs are enlarged and, being too big to fit inside their closed mouths, fold back against the roof of the mouth. When striking, the fangs are erected by a special bone which connects with the jawbone and acts as a lever. Fangs are replaced regularly, each having a replacement directly behind it, which grows into place when the main fang is shed. Venom is pumped through the hollow fangs as the snake strikes, and deaths have even resulted from the reflex action of the jaws when snakes have been carelessly handled soon after death. Their highly toxic venom retains its qualities for long periods, rattlesnake venom having been proved to be dangerous when fifty years old.

Their deeply penetrating fangs, and both the toxic properties and bulk of venom produced by the pit vipers give them almost unequalled killing power. Up to 7cu cm of venom have been collected from the glands of a large fer-de-lance in one milking and over 1,000mm from a rattlesnake. Rattlesnake venom is of the type which breaks down the walls of the blood vessels and the leucocytes, but there are variations in different species. The pygmy, the South American and the western Mexican rattlesnakes are especially dangerous as they also have a well-developed neurotoxic element in their venom which acts on the nervous system, and a special antivenom is needed for treating their bite. The venom of the eastern diamond-back rattlesnake is well known for its necrotising effect on flesh, caused by the digestive enzymes present, and loss of digits or limbs have resulted from its bite. Fer-de-lance venom has perhaps the most spectacular effects on its victim, as it causes haemorrhage and blood oozes from all the orifices of the body.

Despite being such perfect killers, life is by no means idyllic for the pit vipers. Hawks, mongooses and other predators prey upon them, and there are several reported cases of rattlesnakes being killed by their fellow pit vipers, the moccasins. Also in North America the cannibalistic king snake kills and eats them with impunity. Further south the role of cannibal is taken over by the mussurana; it feeds largely upon the plentiful pit vipers of the genus *Bothrops*, of which there are sixteen known species in Brazil alone.

The rattlesnakes occur over a large area of North America from southern Canada to Argentina. The rattle, which functions as a warning only, is composed of horny loosely interlocked segments situated at the end of the tail; it is characteristic of the family and is peculiar to it. Newborn rattlers have a single button on their tails, a segment being added every

Hiding in rodent burrows by day, and occasionally basking in the sun, the diamond-back rattlesnake hunts at night, tracking its stricken prey by gathering scent particles on its flicking tongue

time they shed their skins, which is determined by their growth rate. This is faster in young snakes and up to four rattle segments are added annually, slowing down to one per year as they mature. Rattlesnakes have colonised all manner of terrestrial habitats, from the grasslands and desert to heavy woodland. The sidewinder, a desert species, feeds upon the nocturnal kangaroo rat. It gets its name from the peculiar sideways mode of locomotion, which improves its mobility in open areas of soft, loose sand, where other snakes do not venture. Progression by this method is very swift, and also assists in reducing the snake's body contact with the hot sand, but normally it hides during the day in a rodent burrow. The prairie rattler also hides in rodent burrows by day, usually in the deserted tunnels of the prairie dog, occasionally even alongside occupied ones. Young prairie dogs, and burrowing owls which share the same tunnel systems, are therefore a handy source of food for these rattlesnakes. In the northern part of its range the prairie rattlesnake encounters low temperatures and short summers which result in a slowing down of the growth rate of their young, births occurring every other year instead of annually.

The pit vipers are the most specialised of the nocturnal snakes, adapted for hunting in total darkness. It is as well that the torture masters of old were unaware of their physiology, for they could have devised what would surely have been the ultimate of all tortures. Imagine being confined in a dark room with an adversary highly sensitive to the vibrations caused by movements and low frequency sounds, and an extra sense enabling it to home on the heat waves produced by its terrified companion. Darkness would not hinder it, but man's sight would not be able to compensate for the lack of light, and his senses of touch, smell and hearing would be no defence against a silent, swift reptile equipped with the most efficient venom-injecting mechanism known.

Snakes are relatively silent animals and lack the well-developed communication systems of the mammals and birds, a low hiss being their characteristic sound. Unlike the lizards which have a well-developed sense of hearing, the snakes have no ear drums and the stapes bone is connected to the quadrate bone of the upper jaw instead of to the ear drum as in other reptiles. It is therefore impossible for snakes to receive airborne sounds, and their 'hearing' is in fact a sensitivity to ground vibrations which are transmitted through the bones of the jaw and skull to the inner ear. To investigate snakes' hearing ability, researchers have exposed them to high and low frequency sounds, response only being elicited by the low frequency sounds which resulted in vibration of the cage floor. By swaying their own bodies, snake charmers could quite easily discard their flutes and still entice their dangerous partners to perform.

The boas and pythons have similarly functioning pits to the pit vipers' but theirs are located between the scales of the mouth. The Cuban boa, the largest West Indian species which grows to a length of 14ft, uses its sensory apparatus to remarkable effect in caves, where it is able to locate and catch bats, even on the wing. The most impressive aspect of the boas and pythons is their giant size and supposed constricting powers. The reticulated python and the anaconda share the title of largest snake, reaching a length of about 30ft. Despite tales of 60ft and 80ft-long specimens, skeletons authenticating these lengths have never been produced and the elasticity of snake skins after removal from the body is well known. The African python reaches a maximum length of about 25ft; the largest Australian species, the Amethystine python, has been measured at 20ft, while the boa constrictor, usually thought to be the longest of all snakes, reaches a mere 18ft in length.

Their ability to swallow large animals is also well known, but has been exaggerated beyond reason. No snake could swallow a large ox, and even a 30ft anaconda would experience difficulty in swallowing a man complete with clothes.

Sharing the title of the world's largest snake with the anaconda, the reticulated python of South East Asia reaches a length of about 30ft

Large constrictors do prey upon young tapirs, however, and are capable of overpowering and swallowing good sized crocodilians, deer, adult capybara, and have even taken children on rare occasions. Seizing from ambush is typical hunting behaviour of the larger sedentary constrictors. In addition to their elastic, highly mobile jaws, they have well developed salivary glands and lubricate their prey to assist its passage; their windpipe opening can be extended, enabling them to breathe while large food objects are obstructing the throat. Although snakes only have one lung, the left one being lost to the cause of body elongation, it is thought that they are able to breathe, if pressure of food items cuts off the supply of oxygen to the lung, by means of a section of windpipe which may function like a lung.

The smaller arboreal boas often sleep in low exposed positions during the day, from which it is possible to collect them

Hunting at night, the common krait is responsible for many deaths annually in India. Searching dwellings for rodents, it strikes in response to sleepers' movements

before they are disturbed. The well-camouflaged emerald boa, an attractive green, white and yellow snake, is completely arboreal and has long recurved teeth for securing an effective hold on birds. Sleeping wild specimens can be approached and handled and seldom attempt to bite. Cooke's tree boa, however, is a complete contrast in behaviour, and strikes blindly in every direction when disturbed. Although most boas occur in the tropical forests, one small member of the family, the thickset 3ft-long rosy boa, lives in the desert regions of the south-western United States, its range being completely outside the tropics.

In addition to the snakes with heat-sensory devices, many others, both venomous and harmless species, are nocturnal. Taking the dangerous species first, the family Elapidae contains several well-known examples. The common krait, a highly venomous nocturnal snake, lives upon cold-blooded animals, mainly other snakes, and relies upon sight and vibra-

58

tions when hunting. In India, it kills many people accidentally, mainly as a result of striking in response to a sleeper's movements as it prowls through dwellings at night. The widespread custom of walking barefoot in the East increases man's vulnerability to this small reptile, whose bite, together with that of the larger cobra, causes the death of 15,000 people annually in India alone. Krait venom is more toxic to the human body than cobra venom and the latter is said to be forty times as dangerous as cyanide and twice as toxic as strychnine.

The Viperidae family also contains a fair number of nocturnal species; these too are dangerous to man and beast alike. They are rather sedentary, well-camouflaged snakes which lie in wait until prey comes within their reach. After striking and injecting venom they trail their dying victims by means of their Jacobsen's organs. Fortunately the intensity of vibrations acts as an indication of the size of the approaching animal, and

A puff adder tests its prey with its highly sensitive tongue before feeding. Transferring scent particles to the snake's mouth, the tongue acts as a highly efficient sense organ

it is quite likely that the majority of vipers slip quietly away at the approach of a large animal or man. Three African species, the puff adder, gaboon viper and rhinoceros viper, are the most spectacular, with heavy, boldly marked heads and bodies. They have enormous fangs—2in long in the adult gaboon viper—and their glands hold large quantities of venom. The rhinoceros viper is a specialised forest-dwelling species, however, and is seldom seen, but the widespread puff adder, a denizen of the grasslands and now the cultivated areas, is the most feared snake on the African continent, responsible for many deaths annually. Its quick-acting venom causes gangrene, and rodents succumb in a few seconds. In South East Asia its counterpart is Russell's viper or tic polonga, a boldly spotted snake whose venom has been used to treat haemorrhagic ailments. The night adders, occurring in Africa south of the Sahara, differ from typical adders in the shape of their heads, which are elongated, and extend backwards into the body, unlike those of the other vipers whose glands are concentrated in the heart-shaped flattened head.

Protecting its sensitive eyes, the leopard ground gecko's pupils contract to narrow vertical slits in bright sunlight

Night-active Reptiles

The Colubrid family contains more species than all the other snake families combined, but fortunately most of its members are harmless, and the others only mildly venomous compared with the pit vipers, cobras and their kin. In addition to being adapted for life in practically every type of environment where food is available, many colubrids are also nocturnal. The cat-eyed snake of tropical America and the long-nosed tree snake of South East Asia are typical of the mildly venomous members of the family. The latter shows extreme elongation of the head and has enlarged eyes with horizontal pupils in line with the shape of the head and its markings. These arboreal snakes extend their slender heads, holding the body firm by muscle contraction, to bridge a gap between branches. Binocular vision in snakes is also most pronounced in this species.

The rat snakes of North America are good examples of the nocturnal non-venomous colubrid snakes, being virtually a transitional stage between the terrestrial and arboreal species. Their common name, and in fact their frequently used alternative name of chicken snake, reflect their interest in particular food items, and hence their habitat too, as they are regular frequenters of barns and farmyards. A structural modification of the rat snakes and allied corn snake is the possession of weakly keeled ventral plates, allowing them to climb vertical trees by pressing their bodies against bark projections and hitching upwards with their keeled plates.

LIZARDS

Amongst the lizards, nocturnal habits are less common than in the snakes, and only in one family, the Gekkonidae, are there many night-active species. In fact over three-quarters of the geckos are nocturnal, many having specialised vertical pupils which contract when exposed to bright light. Some have lobed pupils which close into four small holes in a vertical line, each focusing an image on the retina. These occasionally hunt during the day, the advantage of having four images being that in total they stimulate the retina better than one. Others have nocturnal adaptations which are not as readily noticeable, such as the lack of a fovea—focal point on the retina—and alteration in the type of retinal cells. Like the nocturnal frogs and toads, geckos depend upon vision when seeking their prey. Movement attracts them; anything small enough is captured and swallowed, after Jacobsen's organ has tested its edibility. The nocturnal geckos hide during the day in crevices, attics, beneath bark, rocks or similar dark places, and in keeping with their habits are not highly coloured, being mainly brown and grey. In contrast, the geckos active by day, known appropriately as the day geckos, are brightly coloured, one actually being green with red spots.

With the exception of the burrowers, whose ears have degenerated and in some cases are completely fused over, most lizards have a good sense of hearing. This reaches its ultimate

A sluggish venomous lizard of the warm deserts of the United States, the gila monster hunts kangaroo rats at night

Active at night even when the temperature drops to 52°F, New Zealand's tuatara holds the record for reptilian low temperature activity

in the geckos, which have well-developed inner ears. They are noisy reptiles and although little is known of their habits it is thought that voice plays a greater part in their way of life—defence of territory, courtship etc—than in most other lizards. The loud voice of the large tokay gecko, for example, is a well-known sound in dwellings in South East Asia.

In reptiles other than snakes, the lower eyelids are the larger and more movable, in contrast to those of the mammals. In some geckos, as in the burrowing lizards of other families, fusion of the eyelids has occurred, forming a transparent corneal spectacle. While this assists subterranean lizards, the reason for its presence in tree-climbing geckos is not clear. An evolutionary stage before the fusion of the eyelids is the 'window in the lower eyelid' found in some species of skinks.

Many geckos have sharp, curved claws on each toe and toe pads equipped with small hook-like projections, some having a similar arrangement beneath their tails. As these hooks point backwards, geckos must curl their feet upwards from the front to release the hooks, curling them downwards again as the foot meets the surface. This adaptation allows them to walk up smooth vertical surfaces, even glass. The bent-toed gecko of Malaya is one of the exceptions to the rule as it lacks toe pads, but the nature of its toes, which are permanently bent upwards at their base and downwards at the tips, results in its claws being held at right angles, allowing it to climb trees with ease.

The mainly terrestrial banded gecko of the American desert areas also lacks toe pads, but is able to climb over rocks in search of insects by using its claws; while the leaf-fingered gecko of the same region is aided in its rock-climbing insect searches by toe pads which resemble leaves. All the geckos are highly insectivorous and are adapted for life in a variety of habitats. As a result of their house-frequenting habits, they are the most regularly seen nocturnal lizards. Of the many species of house geckos of the genus *Hemidactylus*, several prefer to inhabit dwellings.

The desert regions of the American south-west are the home of some other interesting nocturnal lizards. The well-known gila monster and beaded lizard—the heloderms as they are scientifically known—are active at night, although this is hardly an apt description of their sluggish habits. They are the only known venomous lizards, and hunt by smell or taste, using their Jacobsen's organs to trail prey in the manner of the snakes. They withstand long periods of drought by remaining in their deep burrows until more favourable conditions prevail, and in times of shortage can survive on the fat stored in their large tails. The granite night lizard, also a native of dry southern California, has clear immovable bottom eyelids like the geckos. Hiding from sunshine under slabs of rock and in rock crevices, it emerges at night to hunt for insects, when in turn it is preyed upon by the nocturnal Californian lyre snake.

This seems an appropriate place to include the primitive

tuatara, which is lizard-like in appearance, although not a true lizard. Placed in an order of its own on account of its anatomical differences from the other lizards, it varies mainly in the structure of the skull and the absence of hemipenes in the male. Tuataras are mainly active at night, feeding upon the eggs and small young of the petrels whose burrows they often share. In addition, the rich soil produced by the petrels' droppings attracts and supports a large population of invertebrate life which is palatable to them. They shelter in their burrows during the day, but may appear at the entrances to bask in the sun.

CHAPTER 5

Birds of the Night

Every class of vertebrates, except the birds, contains a number of blind species, and among the birds only the kiwi has such poor sight that it depends more upon its other senses. But whether along the hedgerows or in the bird house of a zoological garden, it is a well-known fact that roosting diurnal birds disturbed at night are unable to find their way back to their perches in the absence of light. The reason is very simple. All diurnal birds have very few rods in the retinas of their eyes, and are thus unable to make use of the low levels of light available after dark. Their large number of cones provides great acuity of sight and colour vision—which, of course, is rather obvious as indicated by the brilliant nuptial garb of many species, and the colour-favouring habits of others, such as the bower birds. As a result, bird keepers who provide artificial light for their charges find it necessary to have a dimmer-system so that the birds are made aware of the approaching day's end and seek perches before the lights are switched off completely.

In contrast to the diurnal birds, the nocturnal species have a preponderance of rods in their retinas and few, if any, cones. Consequently they have no use for colours, either for food selection, attraction of mates or courtship displays. Just like the nocturnal mammals, theirs is a world of black and grey. Despite their highly evolved eyesight, however, nocturnal birds cannot see in complete darkness and, with the exception of those which use sonar to navigate in caves, all birds are helpless in total darkness.

In comparison with the birds which rise with the sun, the nocturnally active species are very few. Although in a large number of birds nocturnal activity may take place at specific times of year, particularly during migration, basically only the members of a few orders are particularly adapted for nocturnal life. These orders are the Strigiformes—the owls; the Caprimulgiformes—nightjars and their kind; the Procellariiformes—the petrels and shearwaters, and the Apterygiformes—the kiwis. Amongst other orders there are isolated examples such as the bat hawk, the two species of nocturnal parrot and the fairy penguin.

The retinas of birds' eyes do not contain the extensive vascularisation of the mammalian retina, and it is believed that nourishment and oxygen are diffused from the pecten through the vitreous humour to the retina. The pecten in birds is a densely pigmented fan-shaped structure which projects into the vitreous humour, its rich blood supply providing the retina with nutrients. It is also thought possible by some authorities that it increases the eye's sensitivity, thereby aiding vision, because of its uneven shadow on the retina; and can therefore be said to have an optical function in addition to a trophic one, ie concerned with nutrition only. The fact that nocturnal birds have the smallest pecten—and the diurnal birds, with the greatest visual requirements, have the largest —adds weight to this theory.

THE OWLS

Most nocturnal birds have well-developed night sight, but none better than the owls with their massive concentration of rods. The owl's eyes are elongated by the greatly enlarged cornea and lens, which allows more light to enter the eye, thereby producing a brighter retinal image. Their rods are very slender and an enormous number are packed into the retina, 56,000 per mm having been counted in the tawny owl. It has been estimated that the owl's eyes are ten times more sensitive than man's; but the suggestion that the location of prey is assisted by the perception of infra-red rays emanating from it in the darkness has long been controversial and is not accepted by most experts. Owls have surprisingly good eyesight during daylight as well, having a number of cones in their retinas in addition to all the rods. The owl's lack of a fovea—the central depressed area of the retina containing cones only—and its much enlarged pupil at night results in

The black-crowned night heron's large eyes indicate nocturnal or crepuscular activity. A widespread species, it frequents swamps and riverbanks, where it searches for small animal life

poor accommodation; that is, it cannot form an image of both distant and near objects because of its inability to adjust the radius of curvature of the lens. This is why owls are short-sighted and always hunt close to the ground. The largest eagle owls have eyes larger than man's, and owls generally have eyes so large that they almost touch each other within the head, being separated only by a thin septum. Their heads simply could not contain larger eyes and, as there is no spare room for muscles either, they cannot move their eyes at all. This could be a problem as their eyes face forwards and therefore have a relatively small field of vision (although it is better binocular vision than in most birds), but this is compensated for by their ability to move the head through 200°. The owls' normal field of vision is no more than 110° and the binocular field a mere 70°, but their head-moving ability enables them to look behind from either side without moving the body, and they can even turn their heads upside down. Most birds in fact are incapable of moving their eyes and rely upon move-ment of the head and neck to extend their field of vision, although none has the owl's range of head movement. So much space is taken up by the eyes that not only is there little room for muscles in birds generally but little room for brains either.

When hunting in almost complete darkness, owls rely upon their highly developed hearing to locate their prey and, unlike other birds and all other vertebrate animals, their ears are asymmetrical. The standard arrangement is, of course, an ear in the same position on each side of the head, hearing there-fore being horizontally symmetrical. This type of ear arrange-ment can pinpoint sounds originating from a horizontal plane, but causes difficulty in quickly locating a sound in the vertical plane. This would be an obvious drawback to owls, which in almost total darkness need to locate the hypersonic sounds of small rodents quickly and accurately. Hence their asym-metrical ears. The owl's ears are of slightly different size and

With 56,000 rods per mm in its retinas, the tawny owl's eyes are many

Opening wide
during darkness,
the short-eared
owl's pupils allow
the maximum
entry of available
light

position on either side, the degree of asymmetry varying in different species according to their activity patterns and therefore their use of hearing as a dominant sense. The area of sensitivity to sounds differs on either side of the head, enabling them to locate their prey, in the vertical as well as the horizontal plane, from a single minute sound.

Owls' outer ear tubes are very wide and some species possess a large erectable operculum or concha on the anterior margin, lying over the orifice and covered by the facial disc feathers. This covering of feathers over their ears reduces wind resistance when they are in flight, but has the obvious disadvantage of reducing the ear's sensitivity and particularly its powers of direction finding. Unlike most birds which have a short feather-covered tube as an external auditory meatus, the outer ear in the owls is such a large cavity that it may extend across almost the whole side of the head. Within the spectrum they are said to have a sensitivity range similar to man's at the high frequency end, whereas most other birds apparently have hearing sensitivity lower than man's range by two octaves at each end of the spectrum. Coupled with this highly developed sense of hearing is a mode of flight which in no way conflicts with it. Their feathering is so soft that their flight is noiseless, unlike the heavy, noisy flight of many birds, such as the heavy wing beat of pigeons and the whirring flight of partridges.

The owls are the nocturnal counterpart of the hawks, falcons and other diurnal birds of prey. Their outer toes are reversible, although they usually point to the side, and their flattened faces form a characteristic facial disc. The females of most species are larger than the males. Surprisingly they vary in their nocturnality, and some are in fact completely diurnal. The well-known tawny owl is a true night owl, whereas the little owl is almost completely diurnal and experiments have proved that it even has sufficient cones for colour vision. The extent of their specialisation is emphasised by the fact that the tawny owl is so lacking in cones that it is blinded by bright

The dull garb of the African eagle owl is typical of the nocturnal owls, and is protective by day. Unlike the diurn birds, owls have no use for bright colours

light, while others, such as the burrowing owl, are known to sun-bathe. Tengmalm's owl further confuses the picture as it is diurnal in the Arctic, where it has little choice during the long summer days, yet is strictly nocturnal in the taiga—the coniferous forests south of the tundra. When they roost in fairly exposed places during the day, owls are able to close their third eyelids independently of each other to protect the eyes from the effects of sun glare.

The nocturnal owls normally conceal themselves during the day in thick vegetation, in tree holes, in old buildings or in underbrush nesting areas. Their protective colouring enables them to roost in fairly open undergrowth without being detected, but when discovered they are usually mobbed so persistently by small birds that they are forced to change their location. That they are able to do this during bright sunlight indicates their ability to see by day, although they have more difficulty selecting a suitable perch and landing than at night. It has been shown that many small species of birds recognise owls by their calls, and do not react to the calls of large owls which would not naturally prey upon them. Larger birds, on the other hand, are alarmed by the calls of large owls native to the same region. The owls, masters of silent flight and noiseless attack, and with sombre plumage, must resort to other means of attracting mates at breeding time. They do this by what are known as audible displays, with often weird and ghostly calls ranging from screeching to hooting, and change to visual displays only when they are at close range.

All the nocturnal birds, except one, are 'predatory' in the sense that they require a diet of animal protein; the only one surviving on an inanimate diet being the oil bird. The owls have a varying food intake from insects for the smaller species, such as the scops owl, to gophers, rabbits and grouse for the eagle owl. In Malaya the eagle owl is said to enter bat caves, but there are apparently no records of it catching bats, its food being listed as rodents, snakes and birds.

Birds of the Night

No other birds of prey are habitually active after dark, but several falcons and the bat hawk are crepuscular, being active for short periods at dusk and dawn. They are all swift fliers, the bat hawks preferring to hunt in the open where they have ample space for manoeuvring and gaining height to drop down on to their prey. They also wait at cave entrances to prey on bats as they are leaving and returning, making use of a food supply that no other bird of prey regularly exploits. They seek only small bats, swallowing these whole while in flight, in order to gain sufficient food in the short time available to them. The bat falcons have similar habits, but return to a perch to devour their prey. These aerial hunters have excellent vision and good powers of accommodation, and in addition to capturing bats are also fast-moving enough to take swifts and cave swiftlets which are occasionally still active during the twilight hours.

NIGHTJARS AND OTHERS

As the swifts and their relatives settle down for the night, a night shift of insect-seekers of another kind make their appearance, or at least make known their presence with a repertoire of the most melancholy calls. In the forests of the New World tropics, the wailing of the potoos breaks the stillness, while in Australia the deep croaking of the frogmouth is a characteristic night sound of the forests. In North America the whip-poor-will repeats its name continuously to the infuriation of campers settling down for the night, and in Europe, Asia and Africa several species of nightjars emit their wide range of penetrating calls only after darkness has fallen. They all belong to the order Caprimulgiformes—all insect-eaters with the exception of the oil bird (see chapter 7), and with few exceptions, completely nocturnal in their activity. Most are solitary, their loud, far-carrying calls acting as a means of communication and of defining their territories, but why they

Relying mainly on its excellent night vision, the European nightjar flies as silently as the owls, and scoops up insects in its large bill

Kiwis have the most highly developed sense of smell of all birds. Their sensitive nostrils, positioned at the end of their long bills, can locate earthworms below the surface

should repeat their doleful calls often to the point of distraction is still a mystery.

The frogmouth has an enormous gape, at first thought to be an adaptation for aerial insect-hawking as in most of the others, but now known to be used for plucking insects off the ground or branches. These birds sleep by day on stout branches in a vertical position, their bills pointing upwards, and are very sluggish if disturbed; although, with their motionless posture and cryptic colouring, they usually escape attention. They are fairly swift fliers at night and have excellent nocturnal vision, their eyes facing forwards as in the owls, giving them the advantage of good binocular vision. In contrast the nightjars normally rest on the ground amongst fallen leaves and branches, but are also difficult to locate as their plumage of mottled brown, grey and rufous effectively camouflages them. Flying silently and erratically at night, rather like

76

the diurnal rollers, they scoop up insects in their large bills, aided by their long and highly tactile rictal bristles. The European nightjars normally occur in drier areas, such as the sandy open woodland and pine forests, whereas the Egyptian nightjar occurs in the desert. The potoos have different feeding habits, swooping like flycatchers from an elevated and exposed perch to collect passing insects. Their bright yellow irides glow at night when reflecting artificial light, and their melancholy cries have aggravated the distraught condition of many travellers lost in the jungle.

KIWIS

New Zealand is the place to see some of the world's more unusual nocturnal birds. A rare parrot active only at night, a miniature penguin, and the flightless kiwis which seek worms with their long beaks, are treasures becoming rarer in the face of the introduced animals which have so plagued that country. The kiwi is the only known living relative of the extinct moa —a flightless bird larger than an ostrich. The kiwi looks less like a bird than any other bird, except perhaps the penguin, and has a penchant for laying eggs weighing over 1lb, or approximately a quarter of its own body weight. Kiwis sleep by day in nest burrows of their own digging, the male apparently sharing the work, making the nest and, like the other living flightless birds such as the emu and rhea, incubating the single egg himself.

The kiwi has very small eyes, and sight is certainly not its most reliable sense. The sense of smell, however, is well developed; its long nasal passages, enlarged olfactory lobes and nostrils situated at the end of its long bill, instead of at the base as in other birds, allow it to seek food below the surface. So well developed are the powers of smell that it walks along with its bill just skimming the surface, occasionally prodding downwards as it smells the earthworms below. Kiwis' bills are extremely sensitive, and together with their facial bristles give them definite tactile advantages for seeking worms. Despite the introduction of many predators into New Zealand, kiwis managed to hold on precariously until legislation protected them and conservationists moved populations to offshore islands. Fortunately their nocturnal habits and concealment during the day in their deep burrows also helped them escape from the many introduced animals. As with the thrushes searching a suburban lawn for worms, the kiwis' sense of hearing is also well developed and possibly aids them in the search for food. They certainly use it in communication, as the males have a high-pitched call and the females a guttural one. Nocturnal activity in a land originally free of predators obviously had high survival value as it allowed exploitation of the worm populations which migrate to the surface at night.

Only two members of the parrot family are completely nocturnal, and one of these—the night parrot of Australia— is possibly already extinct, as it has not been observed in the wild for some years. The fate of the kakapo, the unusual New Zealand nocturnal parrot, is also uncertain as it is now thought to exist only in a small locality of fiordland. If a well-developed **voice and sense of hearing** are considered to be complementary, it is likely that these birds have good hearing because their calls have been associated with the booming of bitterns. They are flightless, yet fortunately for them they hide during the day, like the kiwi, in their self-excavated burrows. Despite its owl-like appearance, the kakapo's eyes are small, and it is likely that the birds' abundant stiff rictal bristles aid their nocturnal searching of low-growing vegetation, and possibly their sense of smell assists them to locate flowers, as they are said to be very fond of nectar. The second of New Zealand's **unusual** parrots—the kaka, similar to the more familiar kea— is also thought to be partly nocturnal as it is regularly heard calling at night.

SHEARWATERS AND PETRELS

The little or fairy penguin also occurs along the southern coast of Australia and is only active on land after dark, nesting in burrows of its own digging or in those of petrels, or in crevices and hollow tree trunks at ground level. Both sexes incubate the egg, taking turns to bring food at night, and guard the chick the following day. As well as during the incubation and fledging period which totals approximately three months, the little penguins are also land-bound, day and night, while they are fasting for about two weeks, and roost in their nests at night during the winter.

The southern coast of Australia is also the home of vast colonies of another nocturnal marine species, the sooty shear-

Visiting land only when nesting, the Manx shearwater times its arrival and departure flights to coincide with the cover of darkness

water or mutton bird, whose gregarious nesting habits have not escaped man's attention and over half a million plump fledglings are collected from their nest burrows annually. The shearwaters belong to an order of birds of which most members other than the albatrosses are nocturnal. They are also known as the tubenoses because of their long tubular nostrils, believed to be related to their well-developed olfactory lobes, thereby endowing them with an excellent sense of smell. These birds are the most expert navigators, converging on isolated islands at breeding time, often from thousands of miles away. They are completely marine, the immature specimens spending most of their first three years at sea, coming ashore on remote islands only when they reach sexual maturity. Only one egg is laid, and an incubation period of up to eight weeks, depending on the species, is then followed by a chick-rearing period of almost twenty weeks.

The shearwaters, and their relatives the petrels, visit land only when nesting and with few exceptions—one being the well-known fulmar—are nocturnal when nesting. Some lay their eggs on cliff ledges, but most avoid their enemies by burrowing, and drill their own holes through the turf if deserted burrows are not available. Several do not build nests, and lay their egg directly on to the moist soil. They arrive at their nesting burrows before daybreak, leaving to feed after nightfall, and take turns to incubate the eggs, the off duty one returning to feed or relieve its mate. All arrivals and departures at the nest site are made at night. Being pelagic they feed either on the surface or dive just below it in search of squid, fish and even carrion, or are truly predatory and feed upon smaller sea birds. As a defence aid at their nesting sites, some are able to eject the evil-smelling oily contents of their stomachs with considerable accuracy for a distance of 5ft.

The allied small storm petrels are also aerial specialists, spending months at sea, hundreds of miles from the nearest land, feeding upon plankton scooped up from the sea as they practically 'walk' over the water, their feet just breaking the surface. Although they are considered to be birds of the night, they spend most of their lives on the wing, seldom settling on the water surface, so day and night are virtually the same to them. Like their larger relatives, they seldom come ashore unless to breed, when they are completely nocturnal, and as their weak legs cannot support their weight they shuffle along using their wings as crutches. They excavate their own tunnels, working at night and sheltering in them by day until they are completed; when navigating towards their tunnel entrances, homecoming birds are greeted by the shrill calls of their mates to direct them towards the correct entrance.

CHAPTER 6

The First Nocturnal Mammals

The most primitive of all the mammals are the monotremes, or egg-layers, whose distribution is restricted to the Australasian zoo-geographic region; but of these the weird duck-billed platypus is diurnal and there is some doubt as to the activity patterns of the spiny ant-eaters, which are the only other members of the order. Some authorities claim they are active in the late afternoon and at night; some say they are crepuscular, while others believe them to be active only during the afternoon, retiring before dusk. Their claim to be nocturnal must therefore be bypassed in favour of the marsupials, about whose nocturnal activity there is no doubt whatsoever. As one offshoot of the mammalian stock emerging from the reptiles produced the monotremes, so another resulted in the pouched implacentals—animals lacking a placenta—and in this great order of primitive mammals practically all the species have nocturnal habits. The first furry marsupials, which hid in the soil and dense foliage to escape from the predatory reptiles, were warm blooded and fed their young upon milk from their own bodies. As the reptiles died out or retreated from the cooling temperatures to the warmer central regions of the world, the mammals evolved to fill suitable niches, but the marsupials managed to hold out only in Australia and South America.

The generally accepted theory is that Australia was colonised by the marsupials and the egg-laying monotremes while it was still connected to the southern land masses. Later flooding of these land bridges and separation of the island from the mainland prevented the evolving larger mammals from following. Thus a population of marsupials was left to evolve in isolation into the forms existing today, although others, such as the giant marsupials, died out for unknown reasons.

The absence of other mammalian competitors allowed the marsupials to branch out and fill all available niches, and there are marsupial cats, dogs, squirrels, ant-eaters and others. Many of these evolved to fill the vacant darkness niche. The presence of representatives of Old World mammals in Australia, such as the rats and bats, is said to indicate that continental drift brought the island close enough to other land masses in later years, and allowed a restricted importation of flying or seaborne higher mammals.

For the marsupials which migrated to South America long ago it was a different story. There, these primitive animals had to compete with the evolving higher forms of life, but several species survived, little changed from their ancestors. Although seemingly stupid animals devoid of all feeling, these primitive marsupials have incredible powers of survival and it is significant that they are all nocturnal and experts in the art of daytime concealment.

THE MARSUPIALS OF AMERICA

In the New World, the opossums range from southern Canada to Argentina, this area actually being the distribution of a single species—the common opossum. All others have a more restricted range, generally within the permanently warm regions. Being implacentals, these naked-tailed opossums have, like all other marsupials, a short gestation period. The young are born in a very immature state, approximately twelve days after conception, the minute 'embryos' making their own way to the abdominal pouch where they seize a nipple and become attached as it swells inside their mouth. Although the females usually have thirteen nipples, they often give birth to more than this number of young, the surplus perishing as there is no question of sharing the teats, as is the case with piglets or puppies. The opossums' first toes are opposable, which assists them when climbing, but they are equally terrestrial in their habits and have colonised developed areas, where they raid garbage dumps at night, apparently attracted by the smell. They are still hunted in many areas for sport or the pot, and in the northern-most parts of their range they hibernate in the winter.

The common opossum gives birth early in the year, often in January or February in the southern-most parts of the USA, later in the north where the winters are longer and

A marsupial, therefore related to the kangaroos and possessing a pouch, Azara's opossum is one of the few New World members of this unique order of primitive mammals

The opossums' habit of feigning death, the result of nervous reaction causing temporary paralysis, led to the popular term 'playing possum'

colder; but their delicate young, scarcely the size of a bee, still have a very precarious journey during unseasonal cold weather. They build nests, gathering grass and leaves and carrying these with their prehensile tails, and while they may be primitive and lacking in intelligence, they are tough and prolific. In wild opossums there is usually a tremendously high percentage of old injuries, particularly healed broken bones. Their habit of 'playing possum'—keeping absolutely still and feigning death—occurs as a result of a nervous reaction to certain stimuli which causes temporary paralysis and has important survival value as a defence aid.

In the tropical regions of the New World, particularly in the humid lowland forests, live a number of interesting opossums, surviving mainly on account of their nocturnal habits and seeking a diet of insects, birds' eggs and fledglings, plus small arboreal reptiles and amphibians. In general, the smaller the species the more insectivorous are its requirements. A high degree of tail prehensility is characteristic of most species, and all have large eyes.

Despite their name, marsupials do not all have a marsupium or pouch; and in some species it is represented by an area of bare skin with small flaps on either side. Even amongst closely related species there is great variation in pouch size and formation, as, for example, in the four-eyed opossums, so named because of the two white spots encircled with black above their eyes. The grey species possesses a pouch, and the brown four-eyed is pouchless, consequently nursing its young in tree nests. Surprisingly, however, the grey four-eyed opossum also builds nests in trees or at ground level and sometimes even in burrows.

The woolly opossum has large eyes with bright orange irides and pinpoint pupils in contraction. Its pouch is merely two flaps of skin at the sides of a mammary swelling on which the nipples are situated. This species is mainly arboreal and, while not as strictly nocturnal as the four-eyed opossums, is still mainly active at night. The smallest species are the mouse opossums, which vary in size from that of a mouse to a small rat. They are also pouchless and mainly arboreal, and when sleeping have the habit of lowering their large ears. They are excellent jumpers, being able to move through the branches after their insect prey with incredible speed. All these opossums have large, naked ears, a well-developed sense of hearing, and a keen sense of smell. Their long tactile facial vibrissae also aid the capture of insects.

Only one marsupial is completely adapted for an aquatic existence, even to the extent of having a muscle around the pouch opening to make it watertight. It is completely nocturnal, which in addition to its choice of habitat means it is

Active only at night, the tropical American woolly opossum lacks a marsupium or pouch. Two small flaps of skin protect its vulnerable new-born young

seldom observed in the wild state. The yapock's dense short waterproof coat is reminiscent of a seal skin; it is truly carnivorous, feeding on crabs, fish and other aquatic animals. It nests in a cavity just above water level, which usually has an exit below water level. The little water opossum is also a good swimmer but does not restrict its activities to water, being found in swamps, woodland and on the pampas of Patagonia, where it is said to nest in armadillo holes. In this species the tail is not fully prehensile and there is no opposability of the thumbs and large toes.

In addition to the opossums of the family Didelphidae in the New World, to which all the above species belong, there is also an obscure group known as the rat opossums living

In Australia, the land of marsupials, the eastern native cat has evolved to fill the niche occupied elsewhere by the small carnivores

A mouse in name only, the desert sminthopsis or marsupial mouse is actually a very aggressive little predator. Sheltering by day in burrows or nests of grass, it bounds around under cover of darkness, searching for insects and larger prey

high in the Andes. Although some are known to be nocturnal, relatively little is known of their habits.

AUSTRALIAN MARSUPIALS

In Australia there is a far wider variety of marsupials, with the members of two families occupying the role of predator. The Dasyuridae marsupials are considered to be nearest to the South American species, which are in turn thought to be similar to their primitive ancestors. In keeping with their predatory habits, they are active, alert animals, with well-developed sight, hearing and smell, and take the place of the cats, dogs, weasels, shrews and other small carnivores of other continents. The smallest dasyures are the marsupial mice, the smallest of these having a body length of barely 2in and feeding solely upon insects. The larger species are the size of a rat and are expert killers of the true rats; birds and lizards and animals considerably larger than themselves fall prey to their aggressive hunting habits. The crest-tailed marsupial mouse is so aggressive that it has been likened to the weasels and is a great plunderer of poultry houses.

Disregarding the so-called marsupial cat, which has yet to be scientifically described, the native cat of Australia is the equivalent of the civet 'cats' and mongooses of Africa and Asia rather than the true cats, and is said by Troughton to have 'the bold intelligence of the carnivores'. Nocturnal in habits, they are able to fold their ears down to blot out sounds during the day, and at night are active against vermin and introduced rabbits. Like the true carnivores, their teeth are adapted for a flesh diet and for catching and tearing. The native cats have granular or smooth palm and sole pads, indicating a basically terrestrial way of life, and their pouches are little more than two shallow folds of skin surrounding the nipples. The larger tiger cats are almost 4ft in length and resemble the martens of the Old World in shape, their serrated foot pads labelling them as arboreal animals. Although capable of killing small wallabies, they normally restrict their prey to smaller mammals, birds and reptiles, and have been known to leap from trees to catch passing birds in flight. Their reflexes and swiftness on the ground enables them to catch rodents and rabbits with ease. Large ears, large eyes and long facial bristles are characteristic of the native cats and the tiger cat, and all are as expert at hunting as their counterparts elsewhere. It is also interesting to find that the Australian equivalent of the cats and dogs of other lands call in much the same way as their equivalent west of the Wallace line. The native cats snarl like the true cats when alarmed or threatened. The stumpy dog-like Tasmanian devil growls and whines just like a small dog, and the native 'wolf' or thylacine also has a dog-like bark.

The Tasmanian devil has a large heavy head, large naked ears and enormously powerful molars. It captures its prey with a short scurrying rush after lying in ambush, and near human habitations has become a nuisance towards domestic fowl and sheep. It is thoroughly nocturnal and sleeps in holes, caves, hollow logs or any sheltered position during the day.

The much larger Tasmanian wolf, now virtually extinct, seeks a rocky lair for the daylight hours, appearing at dusk to prey upon kangaroos, wallabies, birds and spiny ant-eaters. After the fashion of the true wolf, Tasmanian wolves are said to trot relentlessly after their prey until the victim is exhausted, when they make their final rush. This behaviour indicates a keen sense of smell, and no doubt excellent hearing, borne out by their large upstanding ears. The wolves are so remarkably like dogs that it is difficult to believe they have evolved dog-like characteristics and hunting techniques isolated from true dogs. They have a wide gape and dentition, similar to that of the canines, for killing, shearing and grinding. For speed they walk on their toes like dogs, but have also been noted to rise on their hind legs and hop, kangaroo-like, when

Practically extinct on its island home, the Tasmanian devil rushes from ambush to capture passing wallabies and rabbits. At times it is also a scavenger

they were required to move faster than their usual trotting gait. After thirty years of uncertainty about their survival, there have been a few recent sightings of this species. Despite their evolution as a dog-like carnivore, however, they apparently could not compete with the introduced dingo on the Australian mainland, where they died out fairly recently.

Bandicoots

The second family of predatory marsupials bear the name Peramelidae, and are practically as carnivorous as the Dasyuridae. They are all known as bandicoots and look rather like a cross between a small wallaby and a shrew. All the long-nosed species are terrestrial and nocturnal. One group, known as the rabbit bandicoots, burrow in keeping with their name, but the others build nests above the ground. Their pouches open backwards and downwards, and all have fairly large naked or slightly furred ears, reaching their

The ring-tailed phalanger or possum of Australia is thought to rely equally on sight and smell when searching the trees at night for its favourite leaves

maximum length in the rabbit bandicoot. They are basically insectivorous, having the long probing snout of the shrews, but they also eat worms, snails, lizards, small mammals such as mice, and sometimes include plant material in their diet. When foraging for subterranean insects and worms, they often cause damage to gardens and are killed despite being an essential part of the balance of nature. Introduced foxes and dingoes also prey on them and they are hunted by the aborigines for the pot.

The rabbit bandicoots, or bilbies, are the most nocturnal species, and have powerful limbs and foreclaws—ideally suited for excavating the tunnels in which they hide during the day. Their tunnels are unusual in that they are steep and spiralling. These animals are said to have excellent hearing and a keen sense of smell, but poor vision; their greatly elongated ears are folded flat when they are sleeping.

Many of the marsupials have evolved herbivorous feeding

Nocturnal and rarely seen, the Australian sugar glider is the marsupial equivalent of the flying squirrels of other lands

habits, some resembling the squirrels and burrowing rodents of the Old World, others filling the niche occupied elsewhere by the deer, antelope and sheep, although hardly resembling them in form even if in feeding habits and exploitation of a particular habitat. Most of the members of the three families of vegetarian marsupials are nocturnal; the two species in the wombat family and the forty odd members of the possum family are strictly nocturnal, and most of the kangaroos and their kind are, surprisingly, most active at night.

Originally called 'opossums' by Captain Cook because of their likeness to the American opossums—the 'o' has since been dropped to avoid confusion—possums exist on the vegetable matter available in trees. They eat all kinds of plant material from leaves and shoots to fruit and flowers, some even surviving on a basic diet of flower nectar. Like their reptilian ancestors, some still include insects in their diet, but the majority have become almost completely herbivorous.

The ring-tailed possums and the cuscuses are the two largest groups, and being truly arboreal have developed strongly prehensile tails. The former are nest builders and shelter in these by day; when active at night they are thought to rely equally on their powers of vision and sense of smell, the youngsters apparently being able to follow trails made by their parents. Their large upstanding ears indicate well-developed hearing also, and the facial bristles are elongated in most species. The leaf-eating habits of these possums occasionally cause concern in orchards, but they are nowhere near as destructive as the very common and widespread brush-tailed possum. This animal has been a pest in New Zealand since being deliberately introduced for commercial reasons to produce skins for the fur trade. The first introduction was made in 1858, followed by numerous other liberations in both South and North Island. In addition to the 'official' liberations organised by the acclimatisation societies, apparently many trappers made their own releases of breeding groups in areas unoccupied by possums, so increasing their range and providing greater trapping proceeds. They multiplied at a tremendous rate and, despite extensive trapping for skins, soon became a pest of the native flora, which had evolved and flourished in the absence of arboreal herbivorous animals for millions of years. There was a large increase in possums during World War II, and the resultant flood of skins on the market immediately afterwards brought about a slump in prices and only first-quality skins were saleable. Consequently there was a continued increase in the possum population, particularly in areas where the skins were not of good quality generally. There were no natural enemies to control its spread, and none of the introduced predators, primarily brought in to help control the introduced rabbit, were of use in combating the possum.

The cuscuses are mainly residents of the humid forests of New Guinea, occurring in Australia only in a similar habitat on neighbouring Cape York peninsula. Their eyes are very large and yellow-rimmed, with vertically slit pupils in contraction. Apart from their long prehensile tails, they are remarkably similar to the Asiatic slow loris in appearance and in their sluggish movements. They also have a similar omnivorous diet, eating fruit and leaves in addition to insects, birds' eggs and fledglings.

Diverging from the main group of marsupials, the burrowing wombats have taken the form of the rodents, and have even evolved similar incisor teeth. With powerful limbs and shoulders and tough claws, they can burrow into the hardest soil, their incisor teeth allowing them to chew through roots when these bar their progress. Their tunnel systems are extensive, depending greatly on the type of terrain, but may extend for almost fifty yards, with a nesting chamber at the end. The burrowing wombats occasionally feed during the day when food is short and sufficient cannot be gathered at night, but though mainly nocturnal in their activity, like the

The wombat has evolved to look, and act, like a rodent. With large teeth and sturdy shoulders and claws, this true Australian marsupial is built for gnawing and burrowing

kangaroos they are not averse to sunbathing. Their eyes are small and sight is not well developed, but their sense of smell is keen and they follow well-worn trails to their feeding grounds.

Kangaroos and wallabies

Where the kangaroos and wallabies are concerned, nocturnal activity does not necessarily include concealment during the day, as they may sunbathe—except during the hottest hours when they seek shelter beneath bushes and tall grass, some even scooping out nest depressions. The majority of species are in fact either truly nocturnal or crepuscular in their habits and when large herds of kangaroos are shown bounding across the open bushland they have usually been disturbed during their daily rest.

Although basically nocturnal or crepuscular, Bennett's wallabies appear by day when the light is dim. One of the toughest of the larger marsupials, they thrive in temperate climes

The First Nocturnal Mammals

It is believed that the kangaroos' saltatorial, or hopping, mode of locomotion evolved as a result of their efforts, in very early days, to escape from the reptilian predators which sought them as food. Also, for moving amongst the rocky scrub country and tussock grass, hopping would have given them an advantage. In keeping with the hopping, the kangaroo has developed a long, muscular tail which acts as a balancing organ. It also serves as a third limb on which it can rest back, like a tripod, and raise itself up to scan the landscape. Its jumping action evolved in the face of predators which could not employ sustained high-powered chasing, but it was no match for the introduced dingo or the domestic horse, and still less for pursuit by the motor vehicle, over the more open areas.

Due to the depredations of introduced animals—not only predators but competitors such as the rabbit—many species

Escape from the sun is not always the natural complement of nocturnal activity. The red kangaroo may spend several hours sunbathing daily, only to bound quickly away when startled

of marsupial in Australia have become extinct or are in an endangered state. The three species of hare wallaby—found only in New South Wales and now very rare—have been observed to scoop out holes beneath bushes and tall clumps of grass, and even to dig a short open-ended burrow, where a nest was built in which to hide during the day. Burrowing reaches its zenith in the rat kangaroos, which are gregarious and live in community burrows of their own digging. Others, such as the banded hare wallaby, prefer to hide in thickets during the daylight hours, making runs through the densely tangled bush along which the larger predators cannot follow.

In the absence of native sheep and goats in Australia, one group of herbivorous marsupials colonised the cliff faces and rocky outcrops of all the mountain ranges, although they have since been exterminated in the areas most accessible to man. Known as the rock wallabies, they differ only slightly from their lowland counterparts, having a less thickened tail which is used more as a balancing organ than for resting back upon. To aid their prodigious leaps amongst the boulders, their hind feet are padded and have granulated soles. Like the introduced rabbits, they warn each other of danger by stamping and, like all species of large grazing marsupials, they grunt when alarmed.

Not only do the kangaroos and their kind in Australia replace the ruminants, but their evolution along similar lines, to fill the same kind of niche, has resulted in a ruminant-like digestive system. One species which has been extensively studied, the small quokka, or Rottnest Island wallaby, has a digestive system like a sheep, complete with similar cellulose-dissolving bacteria in the sacculated part of the stomach, as in the sheep's rumen. Although kangaroos do not ruminate, they often bring small boluses of food back into the mouth.

The three species of kangaroo—the great grey, the red and the wallaroo—are all nocturnal feeders, and while they are still usually active for some time after daybreak they seek shelter as the sun rises. The wallaroo is said to shelter in caves, where the constant temperature by day, and the gathering of dew-laden herbage at night, enable it to go without drinking for a long period. The futility of destroying kangaroos to make room for sheep is illustrated by the fact that, as converters of herbage into protein, kangaroos are far better equipped than sheep to make use of the coarse grasses available to them, and the amount of meat produced by a kangaroo is greater than that from a sheep. In addition, sheep are choosy and, through overgrazing, destroy certain grasses, allowing others to flourish, and so altering the habitat and its survival value for the native animals.

CHAPTER 7

Navigators in the Dark

Towards the end of the eighteenth century the Italian scientist Spallanzani proved that, to avoid obstacles, bats had to be in full possession of their vocal cords and sense of hearing, but it was not until fairly recently that, due to the efforts of Dr Donald Griffin and his colleagues, man could positively say that bats used echolocation for flying in the dark. By sending out frequencies of up to 150,000 vibrations per second, bats are able to locate obstacles by interpreting the returning echoes, and at such high frequencies the actual length of the sound waves may be no more than a fraction of an inch. Earlier this century vast sums were expended on investigating echolocation as a military aid, and 'sound navigation and ranging'—for which the abbreviation 'sonar' is commonly used—eventually came into use just before World War II.

Despite their powers of avoiding obstacles by echolocation, bats also rely on their kinesthetic or 'muscle' sense, which allows animals generally to move about in darkness in familiar surroundings. Under experimental circumstances, it has been observed that bats accustomed to fixed obstacles in their flying area may bump into newly introduced obstacles, until they grow accustomed to their presence. Experiments have also proved that bats are less able to avoid obstacles when they are tired or have just woken up for their night's activity and are not completely alert.

Surprisingly, all bats do not rely on sonar; to fully understand how their senses are developed and employed it is
necessary to take a quick look at the bats as a group. There are
three basic types—the insectivorous bats, the frugivorous or
fruit-eaters, and the carnivores. Most living bats belong to
the insectivorous group; they are usually rather small and
employ high-pitched calls to navigate at night and to locate
their insect prey which they catch in flight. The fruit-eating
species are usually large and often gregarious, causing damage
to fruit crops in developed areas. Within this group there are
aberrant forms which feed almost exclusively on pollen and
flower nectar, these generally being smaller than the true
fruit-eaters. The carnivorous bats usually feed upon small
mammals, birds and nocturnal reptiles as well as invertebrates. Several are restricted to a diet of fish and crustaceans
which they hook out of the water with their large claws, and
three species of vampire bat live solely on blood.

EXPERTS IN ECHOLOCATION

It follows that animals dependant upon echolocation for survival must have well-developed larynxes and act upon the
returned echoes. Experiments show that bats send out signals
rising from ten per second, when about to take flight, to fifty
per second as they manoeuvre around obstacles. Navigating
difficult obstacles resulted in up to two hundred signals per
second in short bursts. As the duration of these sounds ranges
from one-hundredth to one-thousandth of a second, the speed
with which echoes are received, transmitted to the brain,
their information analysed and a decision flashed to the wing
muscles for action can scarcely be credited. Just as amazing
is the fact that they can distinguish between obstacles and
prey, even when amongst trees or in other areas where there
could be much interference from stronger echoes. Dissection
of the bat's brain has revealed some interesting, yet not alto-

gether surprising information. In the Megachiroptera (mainly large fruit-eaters) the areas of the brain dealing with vision and hearing are comparable in size, whereas in the Microchiroptera (mainly insectivorous bats) there are extensive auditory areas and reduced portions concerned with vision.

Depending upon the species, bats emit sounds through the mouth or nostrils, and in many the elaborate nose leaves are held to be sensory aids connected with sound emission, the sounds being orientated by the leaves. Their calls also vary according to their habits. Some insectivorous species employ ultrasonic sounds of a single pitch, others are able to vary the pitch of their calls, and have been called frequency-modulation bats, examples being the insectivorous North American

Relying on echolocation or 'sonar' to avoid obstacles, the Australian white-striped mastiff bat easily avoids the tangled branches of its forest home

species. In addition to their ultrasonic calls, inaudible to the human ear, bats have other calls, often several types, which are audible to man. All bats are thought to possess echolocation powers, yet some tropical American fruit-eating species of the family Phyllostomidae utter through their nostrils such short-duration soft-sound impulses that they are also known as the whispering bats, their calls having only a fraction of the sound energy of the insectivorous species. Only one genus of the Old World fruit bats—those of the genus *Rousettus*—are known to produce supersonic calls, but there is no questioning the presence of well-developed larynxes in the larger gregarious species at least, as both roosting and feeding times are accompanied by loud squabbling. One member of this last group, the hammer-headed bat, has a greatly enlarged larynx and vocal cords, the former being so well developed that it practically fills the chest cavity, the heart and lungs actually being pushed backwards and to one side to make room for it. Their deep croaking calls—likened by Land and Chapin in their *Bats of the Belgian Congo* (1917) to 'a pondful of noisy woodfrogs, greatly magnified'—is thought to be used for attracting females at breeding time.

From the speed of their calls and the rapidity of their collection on return and passage to the brain for action, it is quite obvious that, in the insectivorous and carnivorous bats at least, the sense of hearing is so acute that they can receive and interpret an echo even while producing a call. Allied to their use of ultrasonics to locate food, compared with the slower-flying fruit bats, are specialised ear structures not present in the others. Their complex ears have an additional hearing aid, the tragus—a lobe positioned inside the front of the ear margin. This varies in shape, being erect and tapering in the North American little brown bats and blunt-ended in the pipistrelles.

Most bats feed by night although a few appear during the early morning and again in the evening before it becomes really dark, and vision obviously plays an important part in the navigation of these species. All bats have functional eyes but, with the exception of the members of the Pteropodidae, they are small and usually almost hidden by fur. Of the Old World fruit bats, only in the members of the genus *Rousettus* are echolocation techniques employed for navigation, the other large-eyed species navigating by sight and locating ripe fruit by smell. Their eyes are particularly adapted for night vision because of the projections from the eyeball which penetrate the retina, increasing the surface area for rods. Despite this night-flying aid, the fruit bats are most often abroad during twilight, and on very dark nights frequently fail to negotiate difficult obstacles such as telegraph wires. The majority, if not all of the bats, do not have cones in their retinas, in keeping with their nocturnal habits, but the

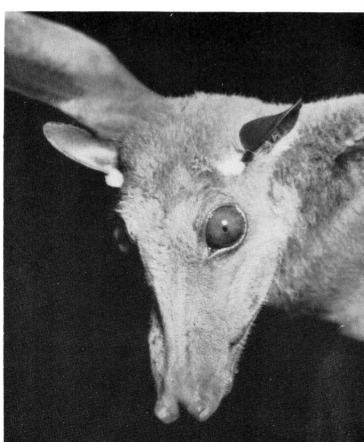

The African epauletted bat's large eyes indicate that sight is the most important sense of this fruit-eating species. The surface area for rods in its retinas is increased by finger-like projections of the eyeball which penetrate the retina

Microchiroptera species are not known to possess the retina-penetrating projections.

BATS ON THE WING

In addition to their powers of voice and acute hearing, bats differ from all other mammals in being the only ones completely adapted for an aerial existence. The framework of their wings is formed by the typical bones of a vertebrate forelimb modified to such a degree that the third finger is enormously elongated and can equal the length of the animal's head and body. The three fingers are almost as long, while the thumb is short and free. These elongated fingers, with the forearm and upper arm, support the web of the wing and have been likened to the ribs of an umbrella. With their extremely flexible joints, bats are able to control speed and direction, and can adjust rapidly to changes in air speed or

Illustrating the extreme modification of the vertebrate forelimb for an aerial existence, the mouse-eared bat in flight clearly shows the upper arm, forearm and elongated fingers which support the wing membrane

air currents. Spreading from the sides of the body, the wing membrane is also attached to, and between, the hindlimbs, often including the tail, where it is known as the interfemoral membrane. Another membrane—the anti-brachial—links the neck with the forearm. Inside the ankle joint, a cartilaginous spur, called the calcar, spreads the tail membrane.

The slender long bones of the bat's arm are hollow and contain marrow; as they are richly supplied with blood vessels and nerves, the flying membranes are highly sensitive and are ideally suited for netting insects in flight. Some species, particularly the horseshoe bats, are even ambushers, as they may seek their prey while hanging upside down, their flexible hip joints allowing them to turn through 360°. When their returning echoes reveal an approaching insect they swoop out to gather it. In the insectivorous species, prey is seldom caught directly with the mouth, the large wing membranes being used almost as scoop nets to enfold insects and bring them within reach of the jaws. Their feeding habits can often be ascertained from the structure of their wings. The high flyers, which seek fast-flying insects, have long, tapering swallow-shaped wings, examples of this group being the sheath-tailed bats and the noctule bat. The low-flying species, like the horseshoe bats which prey on slow-flying moths, have rounded, stubby wings and an almost butterfly-like flight.

Fossil remains have proved that bats looked very similar 50 million years ago to those of the present day. The missing link between the early forms and their ancestors has yet to be discovered. It is likely, however, that they developed from the insectivores long ago, becoming adapted slowly for flight in much the same way as the birds, gradually launching themselves into the air until their webbed forelimbs evolved into the wings they have today.

To enable the forearm to act as a wing, several structural changes have been made in the skeleton to provide a rigid support. The highly developed shoulder girdle is held firm by an enlarged collarbone which extends to the sternum or breastbone, and the ribs are flattened and the vertebral column fused in places. To provide power for the wings, the sternum is ridged as a base for the powerful muscles which pull the wings downwards, while a large thorax holding an enlarged heart and powerful lungs provides the endurance necessary for the long hours of aerial activity.

BATS AT REST

Bats have few enemies and their only regular predator is probably the bat hawk, although other hawks and owls occasionally take the odd specimen. When they are roosting communally and particularly in caves, their strength of numbers undoubtedly affords some protection. But if young or ailing specimens fall to the cave floor, they are at the mercy of the many predators, ranging from cave-dwelling snakes, such as the marbled cave snake, to hordes of cockroaches and other invertebrates. The Cuban boa apparently preys on roosting bats and is also said to be capable of catching them in flight. Other arboreal snakes undoubtedly prey on some tree-roosting species. Predators must be deterred by the obnoxious odour emitted by bats; this is particularly strong in the giant fruit bats, which roost by day in the most exposed positions, making no attempt at all at concealment. Most bats are of sombre coloration, which aids their concealment when roosting, although even the painted bat—the most colourful of them all, and possibly of all mammals—is difficult to locate when roosting amongst fading leaves. Bats roosting in exposed positions obviously do not have much objection to sunlight, although they wrap themselves in their wing membranes, shielding themselves and their young from the heat and fanning their wings to reduce body temperature on very hot, calm days. These are exceptions, however, as the majority of bats seek the shelter

Hanging head downwards,
often from high exposed
branches, the tropical fruit bats
use their wings as sunshades,
and rely on their musky odour
to deter predators

of enclosed resting places, such as caves, crevices, buildings, tree holes and even beneath large leaves.

Caves are the most favoured roosting places of the bats because their constant temperature and humidity, despite external conditions, offers protection from inclement weather. Humidity is essential to them, particularly to the smaller species, as their wing membranes soon become dehydrated if the humidity drops below 85° and they do not have access to water. The temperate-region bats always hibernate in caves of high humidity. Some species have far more unusual day-time hiding places than caves, many making use of plant hideouts. The tropical Asian bamboo bats sleep inside bamboo stems, their flattened heads allowing them to squeeze inside cracks, and suction pads on their feet enabling them to cling to the inside of the bamboo. In Africa the brown leaf-nosed bat has been discovered sleeping in the burrows of the crested porcupine, and some species of African painted bats roost in curled plantain fronds, a habit now adopted by several pipistrelles in banana-growing country. Bird's nests are the favourite day-time resort of other painted or woolly bats, which are often found in the deserted woven nests of the weavers. At least two species—the tent-building bat and *Artibeus cinereus*—cut two angular lines from the edges to the median vein of a geonoma palm frond and roost in the shelter of the two drooping flaps.

BATS AND THEIR PREY

The temperate-zone bats are by nature insectivorous, as they lack the regular supplies of fruit and nectar available to the tropical species. Lack of insects during cold weather causes them to hibernate or migrate; although most hibernate, there

Pipistrelles clinging to the inside of a hollow tree, a favourite roosting place. In the colder parts of their range they also use these secure and somewhat insulated hiding places for their winter hibernation

is a certain amount of local 'migration' between hibernation points during the winter. Migrating bats have been known to cover long distances. the record being held by a pipistrelle which covered 720 miles from Russia to Bulgaria in 1939. As the number of bats ringed and recovered is but a fraction of those migrating annually, it is almost certain that even longer distances are covered. The red and hoary bats of North America, for example, are often seen at sea during their southward migrations. and sometimes even reach Bermuda. Many of the temperate species are known to mate in the autumn, with either delayed implantation or retarded foetal development during hibernation, resulting in spring births.

Blood-suckers
In keeping with their liquid diet, the vampire bats have a short oesophagus and narrow stomach, and their front teeth are specially adapted for scooping out flesh, being slightly recurved and having sharp cutting edges. They fly low on their hunting trips and, like the New World fruit bats, are whisperers, emitting low calls with just a fraction of the sound energy of the insectivorous species. Seeking bare areas of flesh or sparsely furred or feathered regions, the vampire bat alights near its victim, makes its way silently to it and scoops out a small piece of flesh, feeding from the freely flowing blood. An anticoagulant in its saliva ensures a continuing supply of blood until the bat is satiated, but the wound may continue to bleed for several hours afterwards. The blood is ingested by a combination of lapping and sucking—the down-curved edges of the tongue producing a sort of open-bottomed tube which is extended and retracted slowly. Horses, cattle and poultry are regularly attacked, as well as wild animals, but dogs seldom fall victim to the vampires because their sensitivity to high frequency sounds warns them of the bats' approach.

Fish-eaters

The fish-eating bat, *Noctilio*, has been the object of extensive research by Dr Donald Griffin, who is responsible for much of man's knowledge of animal sonar. In a screened flightway in the jungle-clad hills of Trinidad, I watched these amazing bats skimming the shallow water to take small pieces of fish and minnows from the surface. It is believed that as the bat flies just above the surface, head bent downwards, its hanging lower lip directs the clicks down to the water's surface and that echoes are returned from the swim bladders of fish just below the surface. Although this seems a precarious way to make a living, *Noctilio* has survived for a very long time. When prey is located it is hooked with the hind claws, and either eaten in flight or carried to the roost in the bat's cheek pouches.

USE OF SONAR BY BIRDS

Echolocation is also used by birds as a means of navigation inside dark caves when roosting and nesting. The oil bird of tropical America and several Asiatic cave swiftlets emit low frequency sounds, the delay in the echoes' return indicating the distance of the obstacle. The oil bird's frequency range of 6 to 10 kilocycles is much lower than the bat's range of approximately 70 kilocycles, yet they have a remarkable sense of hearing. The intervals between clicks have been measured at 2–3 milliseconds, proving that their rate of auditory response is beyond the capabilities of the human ear.

Oil birds differ ecologically and behaviourally from all other birds; while being closely related to the nightjars, they resemble the owls in many ways. They do not have a crop and food collected for their young is brought back to the nest in the stomach and then regurgitated. Oil birds are the only frugivorous nocturnal birds, as all the other species prey either upon small vertebrates or insects.

Navigators in the Dark

During darkness the oil birds gather the fruit of certain palms and trees of the Lauraceae family, plucking them off in flight with their strong hooked beaks. It is thought that they find their way to the feeding trees and obtain their food by sight, as their large eyes are very light-sensitive, and outside their caves they do not make their echolocating sounds, only harsh cries. The oil birds' keen sense of smell must also help them locate the rather aromatic palm fruits, as their olfactory organs are sufficiently well developed for them to be considered macrosmatic animals, whereas practically all other birds, kiwis excepted, have a poorly developed sense of smell and are said to be microsmatic. Unlike the cave swiftlets which employ echolocation to navigate their dark caves and leave to seek food by daylight, the oil birds, being nocturnal, are never exposed to daylight.

The cave swiftlets' feet have three toes directed forwards and one backwards, and bear sharp recurved claws enabling them to cling to the smallest projection when roosting and nesting in caves. They build their edible nests of a secretion of mucoprotein arising from the sublingual salivary glands, not—as with pigeons—from glands in the proventriculus. Experiments conducted by Tom Harrisson at the Sarawak Museum indicated that their frequency varied from 1.5 to 55 kilocycles, giving six clicks a second—also much lower than bats' frequencies. The echolocating birds apparently do not have the ultrasonic sensitivity of the bats, and use frequencies within man's hearing range. It is quite obvious, however, that although they have a lower hearing range than the bats, speed of response to sounds is far greater than man's.

CHAPTER 8

The Insect-eaters and Their Allies

Although birds are generally considered the world's most efficient controls of the insect hordes, for sheer quantity of intake none can compete with the insectivorous mammals, some of which feed solely on insects of a specific type, and may consume a gallon of ants and termites in one night. Those mammals which are truly insectivorous in their eating habits belong to four orders:

ORDER	HABITS	FEEDING HABITS
Edentata		
Ant-eaters	arboreal	purely insectivorous— ants, termites
Armadillos	terrestrial	insectivorous, carnivorous, scavengers
Sloths	arboreal	herbivorous
Pholidota		
Pangolins	terrestrial/ arboreal	purely insectivorous— mainly ants and termites
Tubulidentata		
Aardvarks	terrestrial	purely insectivorous— mainly ants and termites

114

The Insect-eaters and Their Allies

ORDER	HABITS	FEEDING HABITS
Insectivora		
Solenodons	terrestrial	⎫
Tenrecs	terrestrial	insectivorous,
Hedgehogs	terrestrial	carnivorous,
Moonrats	terrestrial	scavengers,
Shrews	mainly	occasionally
	terrestrial	omnivorous
	some aquatic	⎭
Water shrews	aquatic	insectivorous, carnivorous, piscivorous
Moles	fossorial (see chap 10)	mainly earthworms
Desmans	aquatic	insectivorous carnivorous, piscivorous

The pangolins and the aardvarks are the sole members of their respective orders, whereas the larger orders—the Edentata and Insectivora—contain animals which have deviated from the normal type and have, for example, adopted aquatic habits and feed upon crabs and fish. Others, in complete contrast, have become herbivorous.

The most confusing aspect of the insect-eating mammals is that the only completely insectivorous species do not belong to the order Insectivora, the members of which certainly eat insects but the bulk of their food intake is made up of other forms of animal life. Unlike the ant-eaters, they are not confined to a diet of animals small enough to be gathered on the tongue and drawn through a narrow tubular snout. They are also equipped with lots of teeth, unlike the ant-eaters, and readily accept all manner of small animal food, from young birds, small rodents and earthworms, and are not averse to carrion.

115

THE TAMANDUAS AND SILKY ANT-EATERS

Taking the true insect- or ant-eating mammals first, there is a group of tropical American animals actually known as ant-eaters, although fortunately the two nocturnal forms have additional common names. The tamandua, the largest of these, has such a small mouth opening, barely half-an-inch in diameter, that its choice of food is obviously restricted to small insects such as the ants and termites. Its elongated snout and long tongue have evolved for probing into the nests of these insects, withdrawing those which adhere to the sticky secretion of the salivary glands covering its tongue. The sense of smell is well developed in the ant-eaters, and it is believed

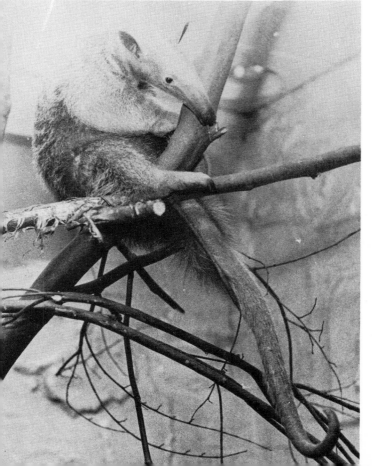

The tamandua has a highly developed sense of smell and a prehensile tail, both of which aid its search for ants and termites in the tropical American forests

that they locate ant and termite colonies by this means. Their eyesight and sense of hearing are said to be poor.

As the tamanduas lack teeth and cannot chew their hard-bodied prey, they have evolved the equivalent of a bird's gizzard, where muscular grinding action reduces the insects. They are completely nocturnal and seldom venture from the trees to the ground, where they are very clumsy in their movements. Tamanduas are inoffensive animals whose only form of defence lies in the strength of their forearms and long claws, particularly in the middle fingers, which are an adaptation for ripping open termite and ant nests, but most useful for repelling attacks by predators. They are also capable of discharging a repulsive odour when alarmed.

The other nocturnal species is the pygmy or silky anteater, also arboreal in habits, but much smaller than the tamandua. Its soft, silky fur is a beautiful burnished gold in colour, and has important protection value when the animal is curled asleep in the foliage by day. It too is inoffensive, but like the tamandua is capable of inflicting a deep wound with its sharp clawed forefeet. In addition to its prehensile tail— one of nature's finest tree-climbing aids—the silky ant-eater has extra-long soles, which are jointed and give a degree of 'opposability', allowing a very firm grip to be gained on branches. It is believed that these animals leave their offspring in tree holes, but a specimen in my possession for some time never once allowed her youngster to become detached from her body.

AARDVARKS AND PANGOLINS

In Africa the role of the ant and termite eater is taken over by two very different animals, the purely terrestrial aardvark and the pangolins, which are both terrestrial and arboreal. The aardvark, the sole representative of its order, is the size of a large pig, its name actually being Afrikaans for pig. It

Large ears indicate excellent hearing, and the aardvark of the African savannahs can apparently locate columns of ants on the march at night

has a long snout and a thick tapering tail, and for such a large animal to feed solely on ants and termites indicates the density of these insects in the tropics. Like the edentates, other than the ant-eaters, their teeth are continually growing and lack enamel; the ants gathered on their long viscous tongues are ground before they enter the stomach. The aardvark's powerful claws give it a capacity for digging unequalled in the animal kingdom, and it is able to burrow out of sight in soft soil far more quickly than a man armed with a shovel.

The aardvark's long tubular ears can be moved independently and are folded backwards to prevent soil entering when the animal is digging. Its sense of hearing is so well developed that it is said to locate ants on the march from the rustling noise made by the column. The tactile hairs on the end of its

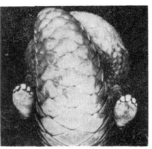

A gluttonous ant-eater, the Asiatic pangolin's small eyes and poorly developed ears are thought to indicate that it relies mainly on its sense of smell to locate food

blunt pig-like muzzle obviously assist food location. Aardvarks normally dig their own burrows, which penetrate several feet into the soil before opening into the enlarged nesting chamber. Sometimes they shelter in termitaries, their thick tough skins providing ample protection from termite bites. They are often gregarious in their nesting habits, and many tunnel in the same area, yet they appear to be solitary hunters, travelling several miles at night to find sufficient food. Aardvarks are powerful animals, their strength lying mainly in their forelimbs which are adapted for burrowing. When attacked by leopards, hunting dogs or lions, they turn on to their backs and slash with all four limbs, and their thick muscular tails. With their acute hearing and surprising swiftness they are quick to regain the safety of their burrow, if attacked while still in its vicinity.

The scaly ant-eaters, or pangolins, could be considered the Old World equivalent of the armadillos, although with the exception of the South African pangolin they climb trees and have prehensile tails. Like some armadillos they can curl into a tight ball, their overlapping sharp-edged scales protecting them from all but the most determined large predators. When extra protection is called for, they emit a discharge from the anal glands. Sheltering in tree holes or in burrows of their own digging, they appear at dusk. From the small size of their eyes, lack of external ears and small tympanic bullae lacking a meatus, it seems apparent that sight and hearing are not their most acute senses. Smell is well developed, however, and they are believed to locate ants and termites by scent. After breaking the nests open with their powerful long-clawed forelimbs, the pangolins secure the insects on their

The hairy armadillo has a keen sense of smell and fairly good hearing. A fast runner, it seeks the safety of its burrow when threatened, plugging the entrance with its body

sticky tongues. It would not be surprising to learn that the sense of taste is also well developed in these animals.

For many years the pangolins were classed with the American ant-eaters because of their close resemblance, particularly their complete lack of teeth. They even have a 'gizzard' type of stomach, whose thick muscular walls grind the hard-bodied insects. It is now thought that these resemblances have merely evolved because of their similar way of life and that they have no affinities with the edentates.

ARMADILLOS

Returning to the order Edentata, the armadillos have a far more varied food intake, even though insects figure prominently in their diet. They are partial to young birds, eggs, amphibians, and snakes, which they have been seen to kill with the sharp edges of their body plates. Even carrion is welcomed, and they burrow beneath carcases to feed and to seek maggots. Armadillos occur mainly in the grasslands, where they are usually solitary creatures. All are terrestrial, with powerful forelimbs for digging, and the majority live in burrows when they are inactive, whatever the time of day. Away from the safety of the burrows, their horny covering protects the upperparts, limbs and tail, but the underparts are mainly soft and unprotected. Folds of skin between their horny plates allow them to bend, and a horny shield also covers the top of the head. Their flight reactions differ only slightly according to their habits and modifications. Rapid running, burrowing, curling into a ball or just withdrawing the limbs into the shell are all employed by the armadillos to escape danger, depending on the type of terrain and the proximity of the predator. The three-toed armadillos are the only ones which can roll into an enclosed ball when danger threatens, but this is no protection against man, and they are consequently becoming rather scarce.

121

Armadillos have a keen sense of smell, and also possess fairly good hearing, but their vision is poor and some have actually blundered into people standing perfectly still. The naked-tailed, hairy, nine-banded and three-banded armadillos have fairly large ears, whereas they are insignificant in size in the other species. The hairy armadillo is one of the smaller species and is only able to withdraw its limbs when attacked, but most predators are unable to make any impression on its horny casing. If attacked after entering its burrow, the hairy armadillo spreads its legs and bends its body to form a plug across the tunnel. It is said to locate food by forcing its head into the soil and then rotating the body so that a hole is formed.

Many authors have written that armadillos cannot climb, but this is certainly untrue of the nine-banded armadillo when provided with suitable footholds. I have observed specimens climbing to the top of a 7ft-high chain-link fence with ease, only the wire roof preventing their escape. Once at the top, however, they were incapable of retracing their footsteps, and dropped back to the ground.

The largest and most powerful species is the giant armadillo which ranges across most of northern South America, and has on many occasions been observed in heavy forest as well as on the grasslands. Adults weigh up to one hundred-weight and have about a hundred small peg-like teeth which are continually growing. They are powerful diggers and have been known to uncover fresh graves to feed on corpses. At the other end of the scale, the pichiciego is the smallest armadillo and unlike all the others. It has soft white hair on its underparts and legs, and the whole dorsal body shell is attached only along the spine and at points above the eyes. The pinkish coloured body-armour extends over the head to the nose and, almost at right angles to the body-shell, another plate at the rear is attached to the pelvis. The small rigid tail protrudes below this plate. Both the eyes and ears are very small and the nostrils point downwards, possibly indicating that the sense of smell plays an important part in food location. When burrowing, the pichiciego supports its raised hind-end on its tail, pushing soil behind it.

SLOTHS

The members of the order Edentata—the arboreal ant-eaters and the terrestrial hunting and scavenging armadillos—vary considerably in appearance and habits, and seem to have little in common. Even the name 'e-den-tate', which literally means 'toothless one', is misleading as only the ant-eaters lack teeth, although all the other edentates lack the front teeth, both incisors and canines. All except the ant-eaters, however, have cheek teeth, both premolars and molars, which lack enamel and resemble each other, except in the two-toed sloths where

Nocturnal habits have certainly assisted the two-toed sloth's survival. Algal growth in its long hair also helps to camouflage it when hanging motionless by day in the tree tops

the premolars resemble canine teeth. It also seems a far cry from the ant-eaters to the leaf-eating sloths, but palaeontological discoveries have shown that they are closely related, the extinct ground sloths being the missing link between the two tree-living types. Both types of sloth—the three-toed and the two-toed—are completely inoffensive vegetarians who feed exclusively on leaves, shoots and flowers found in the highest trees.

The sloth's arms are longer than its legs and, when hanging from horizontal branches, its head is always lower than its backside. It has been repeated many times that sloths spend their whole lives in the same tree, but I for one have never observed a specimen in the same tree for longer than one day. In fact they are usually exceedingly active at night. Escapees from a zoo, where they occupied a large tree surrounded by a wall, travelled up to one mile at night; as overhead pathways were inaccessible to them, they were clever enough to drop from limbs 6ft above ground, crawling to other trees to make their escape even in daylight. On the ground they are virtually helpless, only being able to drag themselves along by their claws into the soil ahead, but they are good swimmers.

Sloths' eyes face forwards and they are from all accounts capable of colour-vision. Their sense of smell is also well developed, but they have poor hearing. Their ears are barely visible beneath the long, shaggy hair. Despite hanging down from the centre of the belly so as to shed water, the hair becomes so saturated during the wet season that algae grow on it. The sloth's fingers and toes are equipped with long, curved claws. All species have three claws on the hind feet and forefeet, with the exception of the two kinds of two-toed sloths, which have only two on their front feet. When active their whole life is spent in an inverted position, although when resting during the day they frequently support their backs in a tree fork while clinging securely with their claws. The claws are used in defence and the sloths' survival has also been aided by their long greenish-grey fur blending with the foliage when they are inactive by day. Their nocturnal habits have also assisted survival, but they do fall prey to the tree-climbing jaguar.

The tenacity of life of the sloths has been commented upon by many authors, and an incident I once witnessed certainly helps to confirm this. A two-toed sloth was dislodged from its branch 80ft high in a giant mora tree while being captured, and landed directly on an exposed buttress root on its head and back. Barely stunned, it soon recovered and was alive and seemingly well nine months later when I last saw it.

Although all mammals are said to be able to maintain a constant blood temperature, the sloths are reptilian-like in the fact that their body temperature resembles that of their surroundings and may vary as much as 10°C. As a result their distribution is restricted to the equatorial zone of the New World. Their body temperature is the lowest recorded for any mammal.

Large ears and a long sensitive snout aid the Cuban solenodon in its search for worms and other small animal life

THE INSECTIVORA

With few exceptions the many members of the order Insectivora live much the same as the armadillos, seeking all manner of insect and small animal food. The order contains many species which have no reliance on light, in fact they are almost all either nocturnal or fossorial, and most of the nocturnal species burrow to hide during the day. The non-fossorial insectivores feed mainly on invertebrates but also accept virtually any animal they can overcome, such as lizards, small snakes, bird's eggs and fledglings, and even carrion. Caged hedgehogs have even shown cannibalistic tendencies towards their dead companions, although it is not known whether this was captive-induced behaviour resulting from dietary deficiencies. The shrews are practically carnivorous and resemble cats and dogs in their aggressive hunting habits. Many insectivores have poison glands for subduing their prey, and a number have a modified body-covering of sharp spines. With very small ears and eyes, they probably depend on their olfactory organs when hunting. Those which are not fossorial, aquatic or burrowing seek cover during the day under debris, or in other animals' burrows.

Some insectivores are considered to resemble their primitive ancestors, and the teeth of all species are of the primitive type. A few aberrant species have diverged to become even more specialised feeders. The potomogale, or otter shrew, is completely aquatic and lives upon crustaceans and fish, and several other species of shrews are semi-aquatic and feed upon frogs and fish in addition to water invertebrates.

Solenodons
The islands of Haiti and Cuba are the home of two of the most unusual insectivores. Known as solenodons, they hide in burrows or hollows during the day, but have little resistance against the introduced domestic animals and mongooses.

They are now exceedingly rare and their prospects for survival are not enhanced by their low reproductive rate, due to the fact that they evolved in a land almost free of predators. They have small eyes and fairly large naked ears. The submaxillary glands of the Haitian species, and it is believed the Cuban species too, discharge toxic saliva at the base of the grooved lower incisors. This has proved fatal to members of the same species after intraspecific fights.

Hedgehogs

Nocturnal and subterranean habits as a means of protection from predators have been adopted by most insectivores, but some receive additional protection from their body-covering of sharp spines. The hedgehogs are one of the most distinctive and easily recognisable of all small mammals. Natives of the Old World only, they are all nocturnal and hide in holes during the day, even in old termite mounds in Africa. In temperate regions they hibernate in these holes during the winter months. The medium-sized upstanding ears of most species are indicative of good hearing, which, with probing snout and well-developed vibrissae, help them to locate worms and other food in the soil.

Surprisingly, hedgehogs are not restricted to wooded areas or regions of rich soil. The long-eared hedgehog lives in the desert regions of north-east Africa to central Asia, and the shorter-eared desert hedgehog has a similar range commencing further westwards. Both dig their own burrows, and in keeping with desert animals generally can survive without water for a considerable time. A captive specimen of the former species once survived for ten weeks without water.

The European hedgehog was introduced into New Zealand from 1870 onwards, although the most successful introduction seems to have been that made at Dunedin in 1885. Early this century hedgehogs made their first appearance in North Island to control garden pests. While being helpful in this

Nocturnal habits and a coat of sharp quills protect the European hedgehog from practically all predators except man

respect and in controlling rabbit and rodent populations, on the debit side they are destructive to the eggs and young of ground-nesting birds, unique to New Zealand, which evolved in the absence of predators.

Tenrecs

On Madagascar and its neighbouring small islands the place of the hedgehog is taken by the tenrecs, which are amazingly like hedgehogs. They are believed to have become established on the island in the Cretaceous period, where in the absence of competitors they showed adaptive radiation and evolved into the many forms which exist today. They filled niches which are occupied in other countries by the hedgehogs and shrews, one species even behaving very much like a mole in its burrowing habits. Like the hedgehogs, tenrecs are able to roll into a ball when danger threatens. Most species hibernate during the winter—from May to October in Madagascar

—by sealing the entrances of their burrows, and during this time have much reduced breathing rate and temperature. They also seal themselves into their chambers when aestivating during the dry season. Their powerful forepaws are also used for digging up the earthworms and vegetable food.

Potomogales

Several insectivores have become adapted for an aquatic existence, and the most interesting of these are the African giant water shrew or potomogale, and the two species of desmans, which belong to the same family as the moles. The potomogale lives along the forest-clad streams of west and central Africa. It resembles an otter, even to the powerful, laterally compressed tail, its only means of propulsion in water. Potomogales have small external ears, very small eyes and a flattened otter-like muzzle surrounded by stiff vibrissae. Small flaps can be closed over the nostrils when they submerge. The entrances to their burrows are below water level, and they hunt at night, mainly for crabs and fish, by scent and possibly touch. A smaller relative—the dwarf African otter shrew—also of the equatorial forest belt is distinguishable by its very fleshy rhinarium compared with the horny one of the potomogale.

Desmans

Only two species of desman are known—the Russian and the Pyrennean—and, although members of the mole family, they look more like long-snouted earless shrews. They are nocturnal burrowers, leaving their tunnels at dusk to hunt for fish, amphibians, molluscs and crustaceans. The laterally flattened, muskrat-like tail of the larger Russian desman and the cylindrical tail of the Pyrennean desman are used for swimming in conjunction with their feet, which are webbed and have hair-fringed edges. Desmans have small eyes, barely noticeable beneath their fur, and from below water level they

129

burrow upwards so that their nesting chamber is above the high water mark. Their long proboscis-like snouts are undoubtedly their main sensory aid, and their musky odour, arising from the scent glands at the base of the tail, make them unpalatable and therefore unattractive to predators.

Shrews

The shrews are found in the Old World only, and although they all have musk glands on the flanks, these are usually more noticeable in the males. They have long facial bristles and often bristles on the tail too. Usually solitary, shrews occasionally live in family parties for some time after litters have reached maturity. They hide under anything large enough to offer shelter during the day, or in holes in the ground, and in Africa often occupy old termite mounds. Most shrews are residents of moist regions, as temperature and the avoidance

Excellent hearing and an acute sense of smell help the aggressive pygmy shrew to locate worms, which it may incapacitate and store for later use

of dehydration are considered to govern their distribution. Little is known of the habits of the few which inhabit arid regions. The discharge from their musk glands is most noticeable during the breeding season, when their acute sense of smell obviously plays an important part in mate location. They also have excellent hearing, but their eyesight is poor and their small eyes are often hidden by fur. The aquatic shrews have stiff bristles on their tails and feet, which act as swimming aids by trapping air bubbles and increasing buoyancy. Shrews are extremely nervous animals, whose heartbeats have been recorded at over 1,000 per minute when alarmed. Their gestation period is very short, varying from two to four weeks depending on the species.

The submaxilliary glands of some shrews secrete a toxic saliva which acts as a nerve poison on their prey. Several quick bites along the length of an earthworm reduce it to a helpless condition in a few seconds. At least one burrowing shrew is known to store incapacitated insects in its burrow. The extremely aggressive nature of the larger species makes them formidable adversaries even where rodents are concerned. In India, the Plague Commission advised against the destruction of the beneficial 6in-long grey musk shrew because its presence in houses kept rats away.

CHAPTER 9

The Desert Burrowers

The moist-skinned amphibians are not the only animals to seek the shelter of the soil during daylight hours; so many other nocturnal creatures take to their burrows at sunrise that a whole volume could barely do justice to them all. Amongst the rodents there are several hundred burrowers, mainly rats and mice, but also such animals as the nocturnal cavies, some of the porcupines and the peculiar sewellel, or mountain beaver. It seems appropriate and advisable in this general work on nocturnal animals to restrict the burrowers to the desert rodents, as representatives of the hordes of rodent burrowers active at night and showing amazing adaptations for survival in a dehydrating environment. Life in their harsh and demanding habitat has resulted in many unique adaptations and habits to reduce the effects of the hot, dry climate. It is as important for these animals as for the amphibians to hide from the direct sunlight, and all the desert rodents included in this chapter have one thing in common—they return to the safety of their sand-girt chambers when the sun rises.

Adaptations for digging and gathering food are, of course, common to them all, but their modifications are also concerned with water storage and the reduction of water loss, plus aiding their movement through loose sand. Having found shelter from the sun and thus overcome one great

hazard of desert life, the finding of water to avoid dehydration—by far the greatest hazard—still remains. This need
not necessarily take the form of a drink from a distant water
hole, however; by eating the succulent parts of cacti and other
desert vegetation, and by licking dew at night, most are able
to gather sufficient fluids for their well being. Apart from
cacti, green vegetable food is usually only available to the
desert-dwellers in spring and early summer, and they must
rely on seeds and cacti for the remainder of the year. They
counteract the scarcity of water by economising in various
ways, such as not sweating or panting and by concentrating
their urine. They are also very partial to insects, which have
a high fluid content. Other desert animals, such as lizards,
are able to cope without actually drinking, because of the
fluid content of their prey, and numerous completely insectivorous birds obtain sufficient liquid from their insect food
and are not known to drink.

Hiding by day is the natural complement of nocturnal
activity where most animals are concerned, and nowhere is
this more important than in taking cover from the blistering
midsummer heat of the desert regions. Many animals which
have successfully colonised the deserts have only managed
because of their nocturnal habits and the microclimate of
their burrows, which protects them during the day. In the
most arid regions there is little existing cover which could
give the degree of shelter necessary for their small bodies,
and the rodents burrow deep into the sand to escape dehydration—and also the intense cold of winter, which is a feature
of the temperate-region deserts. After they have entered their
burrows, many species keep the entrances sealed during the
day in the hottest months by using sand, grasses or whatever
is available locally. Jerboas may even plug their tunnel entrances when they leave at night to forage, particularly when
they have young to protect from predators. Young kangaroo
rats are said to leave the safety of their burrows to search for

food during the day, but adults know better and seldom venture out even on moonlight nights. They are so sensitive to light that in captivity they push into the corners of their quarters to hide.

Most species also seal themselves into their burrows when aestivating during the hottest days of the summer and when hibernating in winter in an attempt to maintain a constant temperature within their burrows. How successful they are at controlling their underground environment has been shown by scientists in the Kara Koum desert of central Asia, who discovered the difference between ground-level air temperature and that just inside a gerbil tunnel could be as much as 30°C. At night the reverse was the case, the ground temperature being much lower than the temperature inside the burrow. In the desert regions of the United States it has been found that an average temperature of 30°C existed in desert rodent burrows despite the intense heat outside, and these animals are known to succumb to temperatures higher than 35°C. Temperature is not the sole consideration, however, as there is less evaporation and consequently higher humidity in the tunnels than at the surface, giving the burrowers added protection from dehydration.

LIFE IN THE BURROWS

The three main groups of desert-dwelling rodents are the jerboas, the kangaroo rats and their relatives, and the gerbils and allied jirds. The type of burrow varies according to the species and their social habits, and even amongst the members of one family there may be considerable variation. In the jerboas, for instance, the hairy-footed species of central Asia are said to construct three types of burrow: a permanent, deep system for breeding; shallow unbranched burrows for nocturnal forays, and deeper burrows for winter dormancy. The desert jerboa of North Africa is gregarious and several may live in one branched burrow which has a number of emergency exits, often ending just short of the surface so that they burst through the sand when threatened. The main tunnel of the long-eared four-toed jerboa of central Asia, on the other hand, is seldom branched, and the small lesser five-toed jerboa of the same region is said to incorporate a narrow vent-hole reaching directly up to the surface.

In the Indian gerbil, which is also found westwards well into Persia, each sex has its own burrowing system, the males having only a single entrance whereas the females have several. As with the desert jerboa of Africa, the exits end just below the surface so that a frightened gerbil breaks through the crust when in flight. The burrow entrances of the desert pocket mice and kangaroo mice are usually located beneath shrubs, and the spiny pocket mice similarly hides its entrances below rocks or logs.

The long, sturdy legs of the desert burrowers support them when they are digging with their fore feet, and to aid burrowing they have tufts of bristles on their feet; these are mainly to facilitate travelling over loose sand, it is true, but also of great help in pushing sand backwards when digging. The desert jerboa has a fold of skin which can be drawn over the nose to prevent sand entering its nostrils when burrowing, and in some species a bony covering to the orbital cavity protects their eyes when they are pushing into the sand. To prevent wind-blown sand sifting into the ear, the true desert-dwelling jerboas have bristles around the ear aperture. Their hind limbs being about four times longer than their front, they are perfectly adapted—like the kangaroo—for a saltatorial or jumping mode of locomotion. Their long tails, often tufted at the end, act as a balancing organ when they are bounding along.

Food storage
The members of the Heteromyidae family—the kangaroo

Large eyes and ears, and sturdy furred hindlimbs are characteristic of the small desert rodents. The gerbil *Gerbillus pyramidalis* is adept at burrowing and traversing loose sand

The epitome of desert dwellers, the kangaroo rat can survive dehydrating temperatures without drinking. Able to manufacture its own fluids, it can actually survive on a diet of dry seeds

rats and mice—have developed food-storage habits to counteract the long periods of drought when food is unobtainable.
For carrying food to their underground storage chambers they
have evolved furlined cheek pouches, which they fill rapidly
with food or nesting material, thus avoiding water loss from
open-mouthed gathering and carrying. They are able to turn
these pouches inside out for cleaning, and when filling them
their small hands move between the supply of food and their
pouches so quickly that the action is blurred to the human
eye. When gathering small mustard seeds, a pocket mouse
was observed to move 3,000 seeds in one hour to its storeroom
some distance away, carrying approximately half a teaspoon-
ful of seeds on each journey. As well as storing seeds of all
kinds in their storerooms, they have been known to gather
the faeces of cotton-tail rabbits. Merriam's kangaroo rat is a
scrounger and seldom gathers its own food from the source,
stealing instead from its larger relatives, whose stores may
contain several buckets-full of seeds and dried grasses, usually
kept in separate piles according to type.

When gathering fresh seeds, the desert rodents bury them
just below the surface, apparently to dry them and prevent
mould occurring in storage. Later the contents of these caches
are transferred to the deeper storage rooms off the main burrow. It has been suggested that some seeds are eventually
taken down to deeper chambers in moist soil, where they swell
with moisture, affording a high fluid intake in times of need.
At least two species—the fat-tailed jerboa of the drier regions
of Turkestan and the pygmy kangaroo rat of western USA—
store food in their tails as a precaution against the possibility
of shortage.

Water conservation

Although most species do not need water, but must have
succulent foods such as cacti in addition to their dry seeds,
many members of the North American family Heteromyidae

can survive indefinitely on a diet of dry grain. The pocket mice, and kangaroo rats and mice, which manufacture their own water from the food they eat, have lived in captivity for several years on dry seeds, and at least one member of this group—Merriam's kangaroo rat—has been the subject of seemingly harsh experiments. 'Kangaroos' in name only because of their shape and bounding habits, they are nocturnal, being able to escape the intense heat of their niche in the United States deserts by retiring to their cool underground chambers during the heat of the day, but they also have what must be the most remarkable physiological adaptation of all the mammals. For a period of fifty-two days in a laboratory, Merriam's kangaroo rats had no water in any form, nor any high fluid-content foods. With a diet of dried grain and a temperature simulating that of their desert home, the test animals carried the same weight of water in their bodies as those which had access to vegetable foods during the same period. How do they do it? The answer lies in their ability to manufacture water from the food they eat. Also they can economise in other ways with a greater degree of efficiency than other forms of desert rodents, which eat cacti and succulents in addition to seeds and can afford to be a little less expert at conserving water. Tests on the laboratory-controlled rats revealed that their excretions contained less water than rodents fed normally; and unlike man and most mammals no moisture was lost by sweating or panting to reduce heat and maintain body temperature.

To appreciate fully how this is achieved, it is necessary to understand the function of the kidneys; these maintain the blood within certain narrow limits of density by excreting excess water or mineral salts which must be passed to the bladder in a concentrated fluid medium. Fluids such as sea water cannot be concentrated by animals' kidneys, because of their denseness, and cause death by increasing the salt content of the blood.

The Desert Burrowers

The kangaroo rat produces its own water from the hydrocarbons it eats and the air it breathes, but it has none to spare and certainly cannot afford to lose water when excreting excess mineral salts. The kangaroo rat reduces water loss in this respect by concentrating its urine to a density greater than that of sea water and seventeen times the osmotic concentration of its own blood. Consequently it could drink sea water without harm. In comparison man can only concentrate his urine to just over four times the density of his blood, and like all other animals would increase the salt content of the blood, hence his thirst if he drank sea water. By producing its own water, the kangaroo rat not only shows an additional means of water conservation, but an immunity to desert drought.

With large spherical lenses and concentric retinas the long-eared jerboa can scan a wide area. Its enormous ears—relatively longer than a hare's —are folded when not in use

CHARACTERISTICS OF THE DESERT RODENTS

An interesting point regarding desert-dwelling nocturnal burrowers is their pale 'desert' coloration. It has been shown that sand coloration is determined genetically and has persisted in captivity through many generations, but it is not clearly understood why nocturnal species, which seldom venture from their burrows during the day, should need protective coloration. It does not apply in every arid area, of course, but depends upon local conditions. Pocket mice in central New Mexico are almost white in areas of practically pure gypsum, while in rugged black lava country not far away they have very dark coats.

Like most rats and mice, the desert species have enlarged, almost spherical lenses and concentric retinas, giving them practically periscopic vision and enabling them to scan a wide area, although with poor powers of accommodation. They rely on this horizon-scanning vision as their main sense of warning, and when searching for food depend more on their sense of smell. Their large black eyes reflect light and appear ruby red at night, just like those of cats and dogs, indicating the presence of a tapetum. In addition to their large eyes, some of the jerboas have enormous ears, relatively longer than the hare's. The desert rodents with small ears tend to have more enlarged tympanic or auditory bullae, containing air of high humidity, and believed to prevent dehydration of the fluid-filled middle ear. These are so enlarged in several species that their heads are out of all proportion to their body size. The sense of smell is thought to be well developed in these rodents as they are able to locate shoots and insect larvae just below the surface. In the Dipodomys family—small rodents of the south-western deserts of the United States—both sexes have a gland situated between their shoulders, the secretion from which is believed to be used for recognition purposes.

The Desert Burrowers

Although desert rodents are known to utter sounds when frightened or injured, their call being a high-pitched and mouse-like squeak, in general they seldom vocalise, but most species drum the ground with their hind feet as non-vocal auditory alarm signals. The lesser Egyptian jerboa, Shaw's jird and the fat-tailed gerbil are known to do this, and the comb-toed jerboa of Turkestan has been heard to drum inside its burrows. It would therefore appear that these animals are sensitive to vibrations passing through soil, and to airborne sounds travelling through their tunnels. Even during the day slight noises made at the entrance of an occupied kangaroo-rat burrow almost invariably result in a rapid response of foot stamping as a warning that the owner is not at all happy about the intrusion. The larger kangaroo rats are unsociable, aggressive animals, fighting furiously with intruders of their own kind, except at mating time, and killing any smaller species which foolishly enter their burrows. In contrast, the small species, such as Merriam's kangaroo rat, which live in colonies, are gentle and seldom fight amongst themselves.

CHAPTER 10

Inhabitants of the Soil

It was no more unusual for the amphibians to colonise the soil than to climb trees in their search for food, and to find shelter from the sun and their enemies—they were, after all, terrestrial animals when they left the water long ago. The rich invertebrate life in the soil was obviously a great attraction to them, as it was to be for the evolving reptilian soil-dwellers and thence the fossorial insectivorous mammals. In search of the subterranean amphibians and lizards, certain snakes became adapted for life underground, and later on many burrowing rodents gradually broke their ties with life above ground and existed permanently in the soil on underground vegetable matter, such as roots, tubers and rhizomes. All these animals evolved unique adaptations to aid their movement in the soil, to enhance their chances of finding food and mates, and to escape the predatory species which followed them into the soil. Many are now capable of leading a normal existence in complete darkness, their other senses being so well developed that the inability to see is no disadvantage. We are only concerned here with the truly fossorial or subterranean animals which seldom, if ever, appear on the surface, as opposed to the burrowers (see chapter 9), which seek shelter underground by day only.

CAECILIANS

The only amphibians adapted to spend their whole lives below ground are the limbless caecilians, which are so primitive they no longer show any traces of a pelvic girdle. Looking rather like overgrown earthworms, some species exceed 3ft in length, with a body diameter of just over 1in. Although some are semi-aquatic and others have an aquatic larval stage, most are moist-soil burrowers seldom seen above the surface. They are occasionally uncovered in gardens and on building sites, and are sometimes washed out by heavy rainfall, as they occur only in the warm lowland regions around the centre of the earth, in areas of high precipitation.

The caecilians are voiceless and have minute eyes which in most species are completely useless because of the degeneration of the muscles and nerves, and the closure of the eye socket by bone growth. They have no visible ears and are said to be capable of hearing sounds of very low frequency only, probably below 250 cycles. How then do the caecilians cope underground without sight or hearing? Their senses of smell and taste are highly evolved. Taste buds are located in the mouth and there are also well-developed Jacobsen's organs which test the edibility of prey before it is swallowed, thereby assisting the cells located inside the nostrils in 'smelling' food items. In addition to the chemically-stimulated senses of smell, caecilians are also aided by their body skin which is an important receptor of chemical sensations. When searching for food, a small but highly tactile tentacle, positioned between the eyes and nose, can be protruded slightly. It is also connected to Jacobsen's organ where the edibility of the source of smell is tested. The tentacle operates independently to the nostrils in detecting food.

Like all other amphibians, the caecilians are carnivorous. Although earthworms are thought to form the bulk of their food, large specimens also include young burrowing snakes in their diet. Their skins are slimy and smooth, and their elongated bodies have resulted in a lengthening of the right lung, and a reduction of the left one.

FOSSORIAL LIZARDS

Towards the end of the Palaeozoic period the reptiles evolved to colonise every available habitat. They developed horny, dry skins and, unlike the amphibians, lungs formed their sole respiratory organs. The survivors of these creatures are insignificant in numbers by comparison, and the amount of fossorial forms is small compared with those species active above ground at all times and those which burrow by day. The only completely subterranean forms are found amongst the snakes and lizards. None exist in the turtle and crocodilian orders. The most primitive fossorial lizards are the amphisbaenas, or worm lizards, which could be considered the reptilian equivalent of the caecilians as they resemble overgrown earthworms, although rarely more than 1ft in length. They are totally

A legless relative of the frogs, so primitive it no longer bears evidence of a pelvic girdle, the caecilian is the only amphibian adapted for a completely fossorial existence

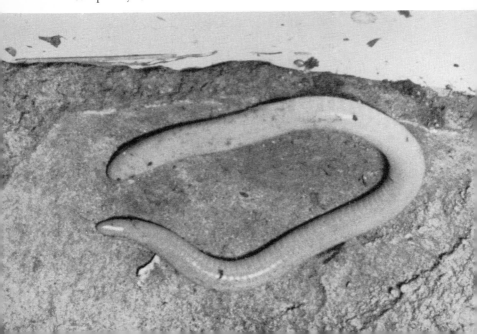

subterranean, but unlike other limbless or nearly limbless lizards which move above ground with a snake-like wriggling, the amphisbaenas move in a direct line. There is still much doubt about their proper place in the reptile line-up. Amphisbaenas eat a variety of small invertebrates, such as cockroaches, worms, beetles and ground-dwelling insect larvae, and often enter termite nests in search of food, their tough hides protecting them from the bites of the termitary guards. They also lay their eggs in the nests of ants and termites. The tropical rainforests are their stronghold, and they are so often mistaken for snakes that two-headed snake worm is a common local name for the spotted amphisbaena of the New World. Its stumpy tail resembles its head, which is shortened and strengthened for pushing through soil.

Loss of limbs or reduction to the point where they are useless appendages occurs as a result of adaptation for a subterranean existence, and at first glance limbless burrowing amphibians and lizards are easily confused with snakes. A family of lizards in Australia fits this category; they resemble snakes more than lizards as their fore limbs are non-existent and their hind limbs have degenerated into completely useless flaps of skin on the sides of their bodies. An example of this group is the flap-footed lizard which, like the other truly fossorial members of the family, has lost its external ear openings, still present in the surface-dwellers. They also have streamlined heads to aid their burrowing.

In some burrowing lizards the movable lower eyelids have evolved into a transparent window to protect the eye from soil and sand particles. Their ears have also become modified for life below ground and are completely covered to protect the ear membrane. The skinks are perfect examples of adaptation for a subterranean existence, and the consequent modifications which enable them to survive. In addition to the structural alterations already mentioned, most of the skinks have vestigial limbs, and a thicker shortened tail which is useful for propelling the body in the absence of legs. Coupled with these adaptations there has been a modification of the shape of the head and jaw in some burrowing forms. As a 'drill-head' to the cylindrical body and muscular tail, the head is extended and tapered, and in some species the lower jaw is countersunk so that it does not extend beyond the outline of the head. An additional modification in some species is the fusion of the head scales to form a protective sheath for their flexible skulls. Skinks have smooth hard scales, which makes earth drilling relatively easy.

Many skinks are completely fossorial and in most of these

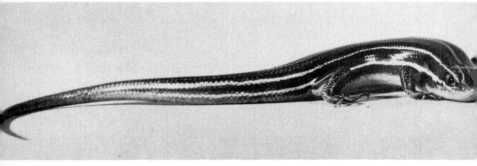

The cylindrical body, muscular tail, pointed head and counter-sunk lower jaw identify the five-lined skink as a burrowing lizard. Its poorly developed limbs are slowly degenerating and in time will become completely useless

the eyes are very much reduced, the eyelids being sturdier and less movable than in the species which just burrow. Some are completely blind because their eyelids have fused. Most skinks have deeply sunken eardrums, external openings of the fossorial species being small and sometimes completely fused over. It is believed that the subterranean skinks 'listen' for sound waves with the whole surface of their heads, picking up the small sounds transmitted through the soil by insects.

The deserts of north Africa, Arabia and Pakistan are the home of the sandfish, an unusual fossorial lizard which has

With shortened heads and blunt tails, sand boas are perfectly adapted for life underground. Eating fossorial lizards and desert rodents they can survive without ever drinking

retained its legs, and has fringes of scales on its toes—an adaptation for running on loose sand—yet despite this they spend practically all their time below the surface. Another unusual family of fossorial lizards is the Anniellidae, the members of which are known simply as legless lizards, the only two known species hailing from California and Lower California. Neither has any trace of limbs, or ear openings, and they have very small eyes and movable eyelids. To aid their progress through sand and sandy soil, their bottom jaws are countersunk, creating a more pointed snout, and their compact heads and smooth scales increase their efficiency. They are said to burrow within a few inches of the surface, locating their insect prey on the surface by touch and their sense of smell.

SUBTERRANEAN SNAKES

Two families of blind snakes—the Typhlopidae and the

147

Leptotyphlopidae—have all the characteristics of the fossorial lower vertebrates. They are cylindrical and have smooth scales; their heads are short and blunt, and their eyes are covered with scales. Their heads have been modified to withstand the pressure placed on the skull when pushing through soil, these modifications being concerned with strengthening the bone. There are many species throughout the warmer regions of the world, some existing mainly on termites and ants, and even laying their eggs in the nests of these insects.

There is still doubt as to whether the Typhlopidae snakes are snakes at all, or if they are in fact legless lizards. Unlike snakes, they are unable to feed on large prey items because of the fixed nature of their jaws, and they are practically toothless. Most species have a sharp spine at the end of the tail which is believed to act as a brace for assisting their movement through the soil. One typhlops is known as the flowerpot snake, now widely scattered throughout the tropics due to its habit of hiding amongst the roots of cultivated plants.

There is no doubt about the classification of the mole vipers and the sand boas, however, nor of the many subterranean forms belonging to the large Colubrid family of snakes. With their shortened heads, which are narrower than the body, and short blunt tail, the sand boas are well adapted for life in the soil, where they feed mainly on fossorial lizards, and on ground-dwelling creatures for which they wait just below the surface. They are also said to enter the burrows of the desert rodents in search of food. Sand boas have tiny eyes, as in their underground environment they obviously seldom have to rely on their powers of vision. Being true desert-dwellers they are able to survive the high temperatures by digging deeper into the sand during the day, and obtain sufficient moisture from the fluids derived from their prey; a captive specimen lived for six years without drinking.

The mole vipers are also completely adapted for subterranean life and, like the other burrowing snakes, have a more cylindrical body than the egigean vipers, plus a narrow head, small tail and reduced eyes. Despite the size of their heads, their fangs are very long—too long in fact to be erected when pushing them over and outside the bottom jaw. Little is known of their hunting habits except that they live mainly on rodents, which they catch in their burrows, and are said to use their fangs in a stabbing fashion.

The reed snakes and spindle snakes of South East Asia and South America respectively occur in the humid forest regions which are rich in fossorial animals. Their rigid skulls and fused head scales aid their capacity for burrowing. Like other completely subterranean snakes, they show a reduction in tail length, as a whip-like tail is of little use to a burrower. Their short and stumpy tails give these snakes the appearance of being two-headed at first glance.

FOSSORIAL MAMMALS

With the exception of the marsupial mole, all the fossorial mammals belong to either the rodent or insectivore orders. The rodents colonised the soil to benefit from the plentiful supply of plant material. Although they may on occasion appear above ground, possibly to gather nesting material, they seldom venture far from their holes, as their specialised senses and adaptations offer little protection from predators above ground.

Marsupial moles

The marsupial mole is undoubtedly the most primitive truly fossorial mammal, and being a true implacental, like the kangaroo, has a pouch, which opens backwards. Varying in colour from white to golden red, they were only discovered in 1888, mainly because their unusual habitat—the dry sandy country of central and southern Australia—was hardly typical mole environment compared with the moist habitat of the true moles. They are completely adapted for subterranean life, however, and so similar to the African golden moles that they were thought originally to have a common ancestor. The marsupial moles are not even remotely related—this being an excellent example of convergent evolution, or adaptation to a particular way of life in an unoccupied niche by unrelated animals on different continents.

The marsupial mole's eyes are vestigial, with no lens and little optic nerve. A horny shield protects its nose and it has no external ears, the ear openings being covered with hair. Fused vertebrae in the neck produce rigidity and aid soil pushing. The limbs and toes are adapted for burrowing, the former being short and sturdy; their digits are equipped with large claws on the third and fourth fingers for scooping soil and flattened nails on the toes which help to throw the soil behind the animal. Marsupial moles burrow a few inches below the surface, the soil falling in after them, leaving no tunnel as with typical moles. Every few yards they surface and shuffle along for a short distance before submerging.

True moles

In contrast to the marsupial mole, the true moles of North America and Eurasia seek damp places and cannot survive for long in dry soil. Their noses are richly supplied with nerves and they have sensory hairs on their hands. The common and western moles are practically blind; their eyelids have almost fused over their eyes due to their almost completely subterranean habits. As they seldom appear on the surface, they have a longer life span—approximately three years—than the hairy-tailed and star-nosed moles, which surface more frequently and are at the mercy of predators. All moles have elongated, cylindrical bodies, and tubular, almost naked nostrils which extend beyond the lower lip. They have short necks and powerful limbs which are furred to the hands and feet, and end in large claws. Their hands turn outwards permanently because of the unusual manner in which the radius articulates with the humerus. Moles' velvet-like flexible fur can lie forwards or backwards, allowing them to move freely in any direction, and it is so dense that whichever way the animal goes soil cannot adhere to it.

After making the first appraisal of soil conditions, a mole burrows by forcing its powerful limbs into the soil and pushing backwards, at the same time forcing its head forwards. It can burrow at a prodigious rate and its body pushes the soil out of the way, making a visible trail when burrowing near the surface. Moles' tunnels are of two kinds, either deep ones, indicated by the typical molehills of soil pushed upwards, or very shallow ones just below the surface. The first are generally for nesting and resting, and the shallow ones for feeding.

The nesting chamber is usually surrounded by an elaborate system of tunnels, no doubt for escape, although it has been suggested they serve the added purpose of allowing aeration of the nest chamber. All fossorial creatures seek maximum tactile sensation at all times, and the moles are no exception, their burrows being little wider than their bodies. Fossorial animals generally are discriminating in the type of soil they live in. Its texture, moistness, chemical composition and aeration are important considerations. It has been proved that moles do not live long in captivity if confined to wooden boxes of soil, yet have a long life span when kept in wire cages filled with soil, especially when these are submerged into the ground.

Moles live mainly upon worms, adding all manner of soil-dwelling invertebrates to their diet whenever they can, but they are not averse to larger prey and have been known to scavenge upon carcases buried in the soil. Normally solitary animals, they apparently do not always fight like demons if they should accidentally enter another's tunnel, as so many authors have reported. The Russian zoologist, Vyazhlinskii, found no such behaviour when studying moles for a long period. They need a very high food intake to counteract their high energy output; they feed irregularly throughout the day and night, gorging themselves and then sleeping for a few hours until the need to feed arises again. They are well aware of the earthworms' habit of migrating to the surface on wet nights and feed just below the surface on such occasions. In very wet areas, they even nest above ground in a pile of soil and vegetation. Experiments by the Danish zoologist, M. Degerböl, with captive moles proved that when there is an overabundance of food they store earthworms after immobilising them. This confirms many previous reports that moles stored food, resulting from the large numbers of worms, with their ends bitten, found in moles' burrows. Strangely, the

The European mole's eyelids are almost fused, but its powerful hands are equipped with sensory hairs, and its nose richly supplied with nerves, so it has no difficulty locating worms beneath the surface

American moles do not store food.

In North America if not the world, the strangest mole is the star-nosed mole, which has a peculiar arrangement of fleshy pink tentacles—twenty-two of them—surrounding its nose. When seeking food, these obviously highly tactile sense organs are in constant motion. Its tunnel systems often lead directly into water, as the star-nosed moles, and in fact many other species, are expert swimmers and may even seek their food along the river bottom. In western North America, surface mounds may indicate the presence of Townsend's mole, surely one of the most energetic of all species; one was observed to form over 300 molehills in seventy-seven days in a small field. In this species the eyes can quite clearly be seen, but their close relatives in eastern North America have much smaller eyes, and smaller ears too, neither of which are visible externally.

Golden moles

The golden moles, which are restricted to Africa south of the Sahara, belong to a different family to the true moles, and although they resemble them they have no visible tail. Their almost metallic coats are usually of a reddish or bronzy hue, but may also be yellowish violet or green, and have dense underfur. All are completely blind, their eyes being vestigial and covered with skin, and their minute ears are barely discernible beneath their fur. The golden mole's muzzle is covered by a smooth leathery pad and the nostrils are protected by a flap of skin. Modification of the chest cavity has provided space for the short, powerful limbs, which are heavily clawed. These moles seek their food below the ground, although two species—the desert golden mole and the giant golden mole—surface at night, their amazing sense of orientation enabling them to find their way back to the tunnel entrance. In addition to their mainly insectivorous diet, several species are known to feed regularly upon skinks

Living most of its life underground the African mole rat ventures to the surface on rare occasions. Sensitive to air movement and vibrations, when alarmed it locates the entrance to its tunnel system with unerring accuracy

and legless lizards, and the desert golden moles have long foreclaws, thought to be an adaptation for catching and killing fossorial lizards. Trevelyan's golden mole, one of the larger species, feeds mainly upon giant worms and often surfaces after dark when searching for them. Despite their small ears the golden moles are very sensitive to vibrations and it is thought that the whole cranium assists in the receipt of vibrations, such as in the fossorial skinks.

Golden moles dig in the manner of the true moles, although relying more on their armoured snouts. Some species, such as the Cape golden mole, burrow just below the surface as the true moles do when seeking food, seldom going so deep that they have to push soil upwards. Others, like some members of the genus *Amblyosus*, normally burrow deeply, sending up large mounds of earth as do the mole rats. They are also said to surround their grass-lined nests with tunnels, and in the event of attack by a mole snake can burst through the cavity wall and escape in any direction. Dewinton's golden

154

mole and others which live in semi-desert regions are more flattened and rather oval in shape compared with the other species, an adaptation thought to aid their progression through loose sand.

SUBTERRANEAN RODENTS

The subterranean rodents are excellent examples of convergent evolution, where unrelated animals have evolved to fill a particular habitat and show several comparable morphological adaptations. The mole rats of the family Bathyergidae in Africa are paralleled by the Spalacidae mole rats and the bamboo rats in Eurasia, and by the pocket gophers of the family Geomyidae in America. They did not originate from the same stock, but their fossorial habits have distinctly affected their structure in similar ways.

Mole rats

The mole rats of the family Bathyergidae are restricted to the grasslands and semi-desert of Africa, south of the Sahara, wherever a supply of underground food is available. The majority are completely fossorial, but some do appear above the surface in search of vegetable food, and like most other rodents they no doubt consume a certain amount of animal matter whenever an opportunity presents itself. Within the family there are several distinctive genera, the members of which have varying distribution. The dune mole rats, for instance, are restricted to the sand dunes of South Africa, whereas the sand rats have a wider distribution from South Africa northwards to Tanzania. The latter apparently appear on the surface fairly often as their remains have been found in owl pellets.

The mole rat's ear openings are very small and surrounded by bare skin; from all accounts hearing is an unreliable sense when they venture above ground. Their sense of orientation is truly amazing, however, for when their tunnel entrances are destroyed they can burrow directly to their chambers. In at least one species—the common mole rat or blesmol—there is a sharp reaction to air movement, vibration and sound. It is believed that a combination of the senses is employed in locating their tunnel entrances, or their underground sanctuaries when the entrances have been destroyed. All the mole rats have enormous incisors, yet both the sand rats and the dune mole rats make more use of their well-developed forelimbs and claws than the others when burrowing. The reason lies in the structure of their jaws, which they are unable to open very wide due to fusion of the mandibles at their symphysis. This affects their feeding habits also as they cannot open their jaws wide enough to carry food to storage chambers. In the Cape mole rat and the blesmol, the incisors play a more important part than the forelimbs when burrowing, being used for cutting through root obstructions and loosening the soil. Their lower incisors can be extended forwards for burrowing and the greater jaw expansion enables them to carry bulbs and tubers to their storerooms. When storing food it is said that they nip the eyes from tubers to prevent them sprouting. Their foreclaws are shorter than those of the sand and dune mole rats, an indication that less reliance is placed on these than on the incisors when burrowing through hard soil.

The mole rats' storage and sleeping chambers are large, and their tunnel systems extensive, undermining the land so much that it is dangerous to walk or ride on horseback over these areas. They are also pests of root crops and there is much controversy as to whether the advantages they bring are sufficient to offset this. The effects of fossorial rodents on the soil have been the object of much experimentation by biologists, and it is generally agreed that soil mixing and aeration are advantageous to the land. From as deep as 15ft down, these industrious burrowers bring soil to the surface, where it is deposited in mounds. It differs from the surface soil in its rich mineral content and lack of humus, and until the mounds have become integrated with the surface soil they bear a different type of flora. Soil movement by these rodents should not be underestimated. In the western United States it has been reliably reported that pocket gophers bring between five and eight tons of soil per acre to the surface. At higher altitudes they render a useful service in holding run-off water in the spring as this sinks into their burrows. Reduction of run-off conserves soil too. In addition to their destructiveness to root crops, however, their extensive burrowing causes erosion, and they also damage ditches and dykes.

Pocket gophers
The pocket gophers are a purely North American family, and all are stocky, neckless rodents superbly adapted for fossorial life. Even their short naked tails are sense organs, being richly supplied with blood vessels and nerves, and are obviously of great assistance when the gophers run backwards in their tunnels. They have small ears and eyes, and their lachrymal or tear glands are highly developed for flushing soil particles from the eyes. The pocket gophers can close their lips behind their large incisors, allowing these to be used for burrowing without soil entering the mouth. Their skulls are heavy, and flattened with wide zygomatic arches; the infra-orbital canal is narrow and protected within the skull from muscle pressure.

Pocket gophers occasionally come to the surface to gather food, and carry large amounts in their extensive external fur-lined cheek pouches which extend back to the shoulders. These pouches are very flexible and, although held back by a muscle, can be turned inside out for cleaning. They live on underground plant material and, like the moles, dig two types of tunnel—long shallow ones for food gathering, and deeper ones for nesting and storing food. The forelimbs are used extensively for digging, in addition to their teeth, and when sufficient soil has been loosened it is held against the chest by the forearms and pushed up into a surface mound.

Diurnal birds of prey capture gophers if they foolishly appear at their tunnel entrances or break the surface during the day, and owls take them at night when they emerge to add vegetation to their diet or make short above-ground migrations. Nor are they secure from predation in their tunnels. Foxes, badgers and skunks dig them out, and weasels and gopher snakes enter their tunnels after them. In Mexico professional trappers—called tuceros—are employed to kill them.

Asian mole rats

Asia also has its own mole rats, although not in the variety which occur in Africa or North America. Members of the Spalacidae family, they occupy most types of soil other than the true desert. Their thick reversible fur is almost mole-like and they also resemble moles externally with their heavy bodies and very short legs, although the legs are not as sturdy as those of the moles and are not turned outwards. Any doubts about their identification are soon removed when their dentition is checked, however, as their projecting incisors clearly indicate membership of the rodent hordes. In addition to their large front teeth, which are used extensively when burrowing, they are also adapted for underground life by virtue of their specialised jaw muscles, so large that they almost fill the orbit, and with their wide horny snouts they are able to ram loosened earth into the walls of their burrows. Allied with the fact that their teeth and heads do the bulk of the excavating, their forelimbs are rather small. The external ears of these mole rats are mere ridges of skin and their muzzles would appear to be their main organ of sense, as on either side of the snout a line of tactile bristles extends as far back as the position of the eye, which unlike all other rodents is covered with skin. Mole rats are solitary creatures, meeting others of their kind only during the breeding season.

Bamboo rats

Another group of Asian 'mole' rats are the bamboo rats, which look rather like enormous moles with the teeth and bare tail of the rodents, yet are adapted for a subterranean existence. Their large orange-coloured incisors are very prominent because the lips are closed behind them, an adaptation for using the teeth when burrowing, without filling the mouth with soil. They also have strong foretoes and claws for digging, and although they occasionally venture to the surface they are slow and cumbersome above ground, their weak eyesight making them vulnerable to many predators. Bamboo rats have the cylindrical, seemingly neckless body characteristic of the subterranean mammals and, as their name suggests, they usually live beneath stands of bamboo, feeding on the roots and shoots.

The bamboo rat's cylindrical, neckless body is the perfect earth borer. Able to close its lips behind its large teeth when burrowing, the bamboo rat prevents soil entering its mouth

Mole voles

All the fossorial rodents we have looked at so far belong to families in which most, if not all, the members are purely subterranean in their habits. There are, however, several very interesting aberrant species occurring in the large rat and mouse family Cricetidae. The mole vole or mole lemming is one of these, and is usually considered to be the most specialised fossorial member of the microtine rodents, which are the members of a large sub-family of rats and mice living in the northern hemisphere. Occurring in central and northern Asia, the mole vole has evolved enormous incisors for burrowing, the upper ones extending forwards as well as downwards. In consequence, with burrowing tools such as these, their foreclaws are small. They have the round head, small eyes and ears and velvet-like coat characteristic of most subterranean mammals, yet despite their habits have the pale 'desert coloration' of the terrestrial arid-land rodents.

Zokors

Another interesting and unusual member of the Cricetidae family is the Asiatic mole rat or zokor, which also lives in central Asia. It is a stumpy animal with powerful limbs and long claws which are turned under when the animal walks. The zokor burrows with its forelimbs, pushing the earth behind with its hind limbs, consequently its muzzle is pointed, unlike the broad pushing muzzle of the 'teeth burrowers'. Zokors have minute, hidden eyes and lack an external ear. Like all subterranean animals, they are sensitive to light despite the smallness of their eyes. They appear on the surface more often than is good for them as it has been reported by biologists working on the Asiatic steppes that these rodents form the bulk of the food intake of the upland buzzard.

CHAPTER 11

The Night Hunters and Their Prey

All animals have enemies, most of them all the time, the fortunate few just at some stage of their lives, and constant alertness and avoidance of predators is the main preoccupation of wild animals, even of most predators themselves. To ensure their continued existence they must be prepared at all times to take flight or to defend themselves or their young. Towards the end of the age of reptiles, the evolving mammals undoubtedly found it advantageous to adopt noctural habits to avoid their diurnal carnivorous ancestors. Their ability to maintain their own temperature without recourse to the sun improved their chances of survival at night. Most mammals, certainly the primitive forms, are colour blind, which is believed to be related to their nocturnal ancestry. As they evolved, many species fell by the wayside and the fossilised remains of numerous kinds are evidence that they were not equipped to compete in the demanding world of nature. Among living mammals the mechanisms concerned with the capture of prey or escape from predators have reached the stage where improvement seems impossible and unnecessary.

The cats and dogs and their close relatives are usually considered the most expert killers. Although some could be included with the arhythmic animals—those which have

irregular periods of activity during both day and night—
many species are crepuscular or almost completely nocturnal
in all their activities. Their hunting methods differ. The
cats are mainly stalkers and ambushers, using little energy
until their final explosive burst of power. Many dogs are
pursuers, practically exhausting themselves by the time their
large prey is within reach, the joint efforts of the pack being
needed to overcome the victim.

CATS AS KILLERS

Hearing, sight and smell are highly developed in the carnivores generally, but never more so than in the cats and dogs,
the latter having without doubt the most expert sense of
smell, whereas the cats have almost unequalled nocturnal
vision. Cats' eyes are the largest of the carnivores and, like
those of the dogs' and hunters' generally, are situated frontally to give binocular vision, hence doubling their chances
of locating their prey. In bright light their pupils are contracted by muscles to protect the lens from damage, and
dilation of the pupils in darkness allows them to make the
most of the available light. In the hunting animals the pupils
in contraction are either vertical slits or are small and circular. In the large and medium-sized cats, and even the lynx,
the pupil contracts to a pinpoint, whereas the reduced opening of the European wild cat, like the domestic cat, is a
perpendicular slit. The presence of the tapetum increases
their powers of vision as it reflects light back through the
retina. The tapetum is virtually a mirror, and replaces the
black light-absorbing layer behind the retina of the diurnal
animals which prevents reflection. Its reflection produces the
night-time 'shine' of cats' eyes when caught in a beam of light.
    The eyes of most carnivores, particularly the cats, are
situated in open orbits, as the socket does not form a complete bony ring around the eye. This allows greater opening

capacity of the jaw, for seizing, holding and bone crushing. Tigers have perhaps the most open orbits and as a result have an extra wide gape. Although closed orbits give the maximum amount of protection from eye injury, in the carnivores a degree of protection is afforded by the bony arch which extends from the orbit to the back of the skull. Strangely, the orbit of the domestic cat is beginning to constrict in the centre and now resembles the dog's orbit, where a bony ligament covers the rear of the eye socket. In times of food shortage or debilitating sickness, cats have a sunken hollow-eyed look due to the loss of fat normally stored in the socket beneath the eye.

As sound-locating organs, the cats' large upright ears are extremely accurate and far more sensitive than human ears to sounds of higher pitch. Cats cannot, however, appreciate the higher pitched sounds of mice, which the mice themselves can hear up to 100,000 cycles per second. At night, particu-

Facing forwards for binocular vision to pinpoint its prey, the rare Spanish lynx's eyes, like those of all cats, are situated in open orbits. This incomplete bony ring around the eye allows greater opening capacity of the jaw for seizing prey

larly in thick forest, it seems likely that hearing and then smell are employed to locate prey and to get within striking range, after which cats no doubt depend on vision. There is probably some variation according to the type of habitat, and in the case of the lion, caracal and similar grassland cats it is likely that sight plays a more important part from the onset of the hunt. In addition to their powers of sight and smell, cats have highly tactile whiskers, whose connected nerves transmit messages swiftly to the brain. When entering narrow spaces the whiskers give an indication of the size of the passage, and the reflexes are so rapid that a living animal just touching one of the facial bristles is quickly seized.

Being skulkers and ambushers, adapted for a short rushing attack after locating their prey from a distance, stealth is the cats' most important asset. Their broad pads act as muffles to soften the tread, and they appear to glide along as they move the limbs on either side of their body together, so that their hind feet fall into the tracks left by the front feet— leaving a single irregular track. Employing the digitigrade method of walking—on the toes—increases speed and agility as it raises the body forward and gives impetus for the final terrifying rush. A cat's tracks in sand or snow, even those of a domestic cat, show an impression of the toes and the large pad immediately behind them. The heel does not touch the ground.

The cats' hunting methods vary slightly in their execution. Co-operation between the only gregarious hunters—the lions —enables one of them to lie in wait at a strategic point while the others drive the terrified quarry towards the ambusher. The solitary hunters may either, like the leopard and puma, leap down on to their victims from an overhead branch or rock ledge, or, like the tiger, lie in ambush beside a jungle track or stealthily approach their usually unsuspecting prey until close enough to attack. In all cases the success of the mission depends upon the final assault, and the outburst of energy needed for this is unequalled in the animal kingdom. All available power is mobilised by adrenalin circulating in the blood stream, causing a sudden flash of oxidation in the nervous system and the brain. Stimulation of the adrenal glands, nervous system and brain increases respiration, heartbeat and the circulatory rate of the blood, sending sugars into the blood stream from the liver. The gorging of large quantities of fresh meat and blood immediately afterwards gives the cats full benefit of the protein and energy content of their kill and allays the exhaustion caused by their explosive outburst. Their most frequent method of killing, where small prey is concerned, is a bite in the nape of the neck; driving their long canines into the hind brain or severing the spinal cord results in instant death. They tackle a large prey animal, such as buffalo, by leaping on to its back and dragging it to the ground, holding the muzzle or throat until it succumbs. Evolution has resulted in the perfection of these animals as killers, and they have evolved hunting mechanisms only, having no needless adaptations for fleeing, like the hunted. The large cats, at least healthy adult specimens, have no natural enemies.

Although most cats are nocturnal in their habits, they are not necessarily skulkers during the daylight hours, and may sunbathe or rest in the shade in full view. Sometimes they also hunt by day, which is usually an indication that they were unsuccessful the previous night. The lion seems to be able to withstand the effects of direct sunlight better than the tiger or jaguar—both basically forest-dwellers—and can be seen moving between shady resting spots in full sunlight. The smaller cats seek secluded resting places and are seldom seen unless flushed out. The serval, for example, usually hides in reed beds and marshes by day, emerging at night to hunt small mammals and birds. Another large-eared small cat, the caracal, prefers to hide among boulders in hilly, dry country and, like the leopard and tiger, stores food when its kill is larger than it can consume at one time. All thoughts of concealment are apparently thrown to the winds at night when, from the deep-throated roar of the lion or tiger to the caterwauling of domestic toms on rooftops, cats everywhere indicate their whereabouts. Attraction of mates or defence of territory are not the only reasons for night vocalising, however, as the lions communicate in this manner when hunting and, with their roaring, drive an intended victim towards their colleague lying in ambush.

DOGS IN PURSUIT

Prey discovery by the dogs at night is most frequently by smell, and they rely upon this well-developed sense and

The African leopard is mainly a nocturnal hunter, spending the daylight hours resting in the shade, often in a tree. Occasionally it hunts by day, an indication, it is thought, that the previous night's hunt was unsuccessful

Having less specialised canines, and therefore killing power, than a cat
of equal size, the coyote is forced to live mainly on rodents and other
small animals which it can easily overpower

steady pursuit to bring them close to their victim, either
individually when they are seeking small prey, or communally when chasing larger animals. It would be pointless, of
course, for a pack of wild dogs to expend energy upon a small
animal as the reward would be insufficient to replace the
energy lost on the hunt. The pack species are strategic hunters which depend upon the exhaustion of their prey to
bring it within their reach. Once they have singled out their
intended victim from a herd, their manner of chasing varies
slightly. Wolves may employ the relay technique, a fresh dog
taking over the lead for a short while and setting the pace,
then dropping back for a fresher individual. Cape hunting

Scavengers both; well-developed senses of smell and hearing bring a side-striped jackal to the scene of a kill, startling the vultures whose keen eyesight gave them first chance. When given the opportunity jackals kill their own food, especially domestic animals

Spotted hyenas quarrel over the remains of a gazelle on Tanzania's Serengeti plains. Seldom abroad by day, hyenas have also recently been proved to be expert pack hunters of young and sick animals

dogs take advantage of their victim's inclination to circle back to the herd; they try to outflank it and reduce the distance covered, never diverting their attention from it even if another animal crosses the trail. Unlike the cats, dogs did not evolve into individuals capable of tackling the large herbivores on their own, but broke off into pack killers.

Dogs have less specialised canines than the cats, and therefore less killing power than cats of equal size. By leaping and tearing at the hindquarters and belly of their victim, or wherever an opportunity presents itself if their quarry turns at bay, dogs eventually kill through their combined efforts, causing many minor wounds. After a tiring pursuit, where large prey animals are concerned, a solitary dog would probably be too exhausted to make sufficient damaging wounds on its own.

All the members of the dog family have a well-developed sense of smell and sometimes, when hunting with muzzles close to the ground, become so absorbed in their task that they completely miss other prey in clear view. The ability of domestic dogs, through their remarkable sense of smell, to detect variations in the activity of man's sweat glands due to illness or emotion, is well known. They also have excellent hearing, as witnessed by the large ears of many nocturnal species. The fennec fox, for instance, has enormous ears out of all proportion to its size, but it is not a pursuer, and hunts by digging its small mammal and insect prey from the ground, burrowing into the sand to rest during the day. Unlike the fennec fox, the very similar bat-eared fox captures small rodents by pursuit rather than burrowing. In the grasslands of temperate South America, the long-eared maned wolf is a solitary hunter, capturing prey after a chase or by digging them out, aided by its elongated canine teeth. Like the cats, the members of the Canidae family have well-developed eyesight and their pupils contract during the day to protect the lens from sunlight. The contracted pupils vary in shape; the foxes have a vertical slit, except for the Arctic and fennec foxes, whose pupils are circular like those of jackals, dogs and wolves.

The long-eared jackals and the hyenas were always thought to be scavengers. It has been known for some time that jackals take lambs, calves and poultry, and will attack even larger domestic animals unable to defend themselves. Hyenas have also taken domestic animals and even the occasional human. All these were considered easy prey for a scavenger. Recent studies by Kruuk in the Serengeti grasslands and the Ngorongoro crater, however, have proved conclusively that hyenas—in this case spotted hyenas—are well able to tackle and kill large wild animals. Hunting like the wild dogs, in packs of up to thirty, they leap at the flanks of their prey until it succumbs. So successful are they in areas such as the Ngorongoro crater, where lack of cover reduces the lions' effectiveness as killers, that the lions themselves rely more on scavenging hyenas' kills than the reverse.

BLOODTHIRSTY HUNTERS

Of course, the order Carnivora contains species other than the cats, dogs and hyenas; many of the almost 200 species comprising the families Procyonidae—the raccoons and their kind; Mustelidae—the weasels, and Viverridae—the mongooses and civets, are nocturnal in their habits. A large number of these, such as the skunks, palm civets, binturong, badgers and red panda, are omnivorous and manage quite well on a diet of fruit and small animals, and even carrion; 'night-seekers' would be a more appropriate name for them. The most expert hunters in this group of carnivores are the weasels and polecats, whose prefix 'bloodthirsty' has been well earned.

Both weasels and polecats can climb, although adapted for a terrestrial existence and particularly for entering rodent

A persistent hunter, the polecat makes short work of conveniently confined prey, such as domestic poultry, killing far more than it could possibly eat

burrows after their prey. Their streamlined form—of narrow head, elongated body, small ears and short legs—enables them to enter the smallest rat burrows. They are persistent hunters, capable of killing animals far larger than themselves, and their victim is usually so terrified after a long chase that it makes little attempt to resist the final attack. They are solitary creatures and always hunt alone. Their depredations of confined domestic animals bear all the trademarks of a massacre; they will kill all the members of a poultry coop, merely eating the brains or drinking blood from a few birds to satisfy their hunger. The importation of the polecat into New Zealand to control the introduced rabbits was one of the earliest forms of biological control, at least of introduced pests. Reports on the efficacy of this move have apparently been rather conflicting; in some areas there

171

was a decrease in the number of rabbits, yet in others an apparent increase. The 'fitches', as these polecats are known, were predatory upon native ground-dwelling birds, which had evolved in the complete absence of predators, and had little defence against them.

THE PREY OF THE PREDATORS

The bulk of the cats and dogs—except the very small species which prey on rodents—feed on terrestrial herbivores, particularly the hooved mammals, which are the most abundant source of protein on earth. The hooved herbivores are primary animals making use of plant material which they convert into meat. Many are gregarious as there are certain advantages to living in a group, and line breeding from the fittest male in the herd ensures continuation and improvement of the senses and the survival of the fittest individuals. By preying upon the weakest and oldest specimens, predators assist in maintaining the vigour of the herd, and even new-born young must be able to keep pace with the herd soon after birth. Competition amongst males results in the strongest and ablest member of the group becoming the herd leader until he is deposed and, by improving their own stock through breeding from the finest male, the herbivores also unwittingly improve the quality of their enemies by making it more difficult for them to make a kill.

The nocturnal herbivores, mainly antelope and deer, are almost equally dependant upon hearing, smell and sight, and naturally have larger and more rounded brains than animals which are mainly dependant upon one sense. Their impulses trigger off the rapid muscular action needed to propel them into flight, and it is obvious from studying the head of a deer or antelope that the large ears, nostrils and eyes are adapted for the receipt of information, and that it is a defensive, not an offensive, animal. The cats are not restricted to

A denizen of the forests, the crowned duiker's large eyes, ears and nostrils are characteristic of the prey animals, always alert for danger. Eyes set on the sides of its head provide monocular vision covering a wide field

feeding upon nocturnal ruminants, but also have access to the many diurnal species which are resting at night. These must be more susceptible to attack from nocturnal predators which are far more adapted for night hunting than the herbivores are for escape at night.

Although the complicated stomach of the ruminants aids digestion it is also considered an adaptation for the avoidance of predators. Without needing to masticate its food as it eats, the ruminant can ingest a large amount in a short time, exposing itself for no longer than necessary. It can then ruminate under cover in comparative safety. The main components of leaves and grasses are cellulose and carbohydrates, which man cannot digest, yet from them the herbivorous animals are equipped to derive all the elements necessary for life. The mass of hastily taken vegetable matter enters the rumen, or first stomach, where the cellulose content is hydrolysed by bacteria, as digestive enzymes are not secreted into this chamber. The softened food is then returned to the mouth in the form of boluses, and is chewed again and mixed with salivary juices. When swallowed for the second time it goes straight to the reticulum, or second stomach chamber. From there it passes to the manyplies, or third chamber, and then enters the obamasum, or fourth chamber, where digestion actually takes place.

In Africa the many large species of ruminants, such as the eland, hartebeest, wildebeest and buffalo, comprise major food items for the lion; while in South East Asia the sambar, axis deer, nilghai and the large wild cattle—gaur, gayal and banteng—are taken by the tiger. Leopards on both continents prey on the smaller antelope and deer, wild sheep and goats, and the pack hunting dogs are more than a match for elk, moose, caribou, antelope and wildebeest, and even large carnivores too old or injured to protect themselves.

SURVIVAL OF THE HUNTED

The eyes of the prey animals are normally set on the sides of their heads, giving a wider field of vision—frontally, sideways and even to the rear. Their sight is monocular, each eye having a separate field, and their pupils are horizontal and widely oval in shape, for scanning a wide area. In the ruminants and horses, they remain in a horizontal position despite head movement. When a predator enters their field of flight, speed is the most important means of escape. The distance which prey animals allow between themselves and predators depends upon the species involved and their experiences of attack in the past.

Flight is stimulated by the sudden oxidation of the nerve and brain cells; this is increased by the rush of adrenalin, resulting in an immediate speeding up of the heartbeat, circulatory rate, respiration and muscular action. The furious rate of metabolism is evident externally by the flared nostrils, staring eyes and trembling body and legs. The prey animals do not live in constant fear, however; between their outbursts of flight energy and sudden alarm, they are able to forget the dramatic near-escapes and live at peace, although ever alert. The African grassland herbivores, for instance, live in close proximity to lions, and while they are always alert they do not appear to be constantly in fear of the predators, showing no signs of nervousness until lions come within their flight distance. The arrival of Cape hunting dogs, however, produces a far different and more startling effect, the prey, apparently being well aware of their tenacity and persistence, quickly vacate the area.

Ruminants belong to the large and varied order Artiodactyla. It also contains the nocturnal hippopotamus and the swine, of which most are completely nocturnal. They are all built for flight, their legs adapted for carrying the body, usually at high speed, away from the source of danger. They walk on their toes, shod with hard, horny hooves—the only defence which antlerless and hornless females have against predators. Even the largest ruminants, such as the gaur and bison, are surprisingly nimble, and the forest-dwelling hooved mammals, whatever their size, are capable of moving with amazing stealth through dense undergrowth.

Horns and antlers, situated as they are frontally, protecting the head and neck, are some defence against predators. By sweeping them sideways, deer may ward off a solitary leopard, but even the largest species are usually unable to cope with the pack hunting dogs when brought to bay. By distracting the victim's attention and attacking from all directions, the persistent pack soon wear down and eventually hamstring or emasculate even the most determined stag.

Concealment from the sun is equally important to some

What better way to escape from the sun than submerging in deep water. Hippos leave their cool retreats at dusk to feed in the grasslands bordering Africa's lakes and rivers

forest-dwellers as concealment from predators; some South American deer—the pudu and several brockets, for example —hide in the thickest parts of the forest. Exposure to sunlight without access to shelter has been known to be fatal to the pudu within three hours. Others are less susceptible to sunlight but, even so, hide in the densest parts of the habitat during the day. The sitatunga of central Africa hides in swamps by day, often in water several feet deep, and emerges to feed on its edges at night. The bushbucks occur in dense cover along the edges of swamps and at night venture out into the more open surrounding areas to graze. The small African dikdiks live in drier scrub areas, hiding by day in rocky outcrops and dense scrub. They are said to favour slightly elevated feeding areas where they can watch for leopards and jackals, and dart back to the safety of thick scrub when alarmed. In the drier parts of their range they can apparently survive without water for several months, managing with the fluids contained in their herbivorous food. Another genus of small nocturnal antelope—the duikers— are mainly denizens of the forest, through which they are able to dash far faster than any predator, fleetness of foot being their main protection. Their counterparts in Asia are the mouse deer, secretive forest-loving animals which are sought as food by small cats, large snakes, and many other predators. They lack horns and have elongated upper canine teeth, but their main means of protection is to run for their lives and in dense jungle only the ambushers can successfully hunt them.

The nocturnal ruminants communicate by whistling and barking; roaring to attract hinds when rutting, and by foot stamping; in addition to these auditory signals, scent-marking is employed by many. The discharge of glands below the eyes, on the hind legs and between the hooves is used to demarcate territory and to attract mates during the breeding season. The glands are particularly useful to the forest-

177

dwellers and the gregarious nocturnal species, where sight is less well developed than the sense of smell, and help to maintain an animal's familiarity with its environment. As the range of visibility of the night-active ruminants is necessarily short, sight is not as well developed as in the diurnal species, especially those which dwell in more open areas, and more reliance must be placed on smell and hearing.

Since the advent of firearms and the increased killing of wild animals, there has been a noticeable lengthening of flight distances as animals learned by experience that man could kill from a distance. Another development has been that some originally crepuscular species have learned that it is safer to adopt nocturnal habits in regions where they have been continually harassed by hunters at dusk and dawn. The Indian hog deer is one of these, and is now almost completely nocturnal in some parts of its range due to human persecution. Fortunately its extreme wariness is accentuated and complemented by its acute senses of hearing and smell. Also in India the wild boar, normally a crepuscular animal, is now nocturnal in areas where it has been over-hunted, which is practically everywhere. In Africa the wart hog is another enforced nocturnal creature where it has been excessively persecuted; as this species was the only truly diurnal pig, it is true to say that all wild pigs are now mainly nocturnal in their habits, the majority of them being strictly so. Like their close relatives the ruminants, wild swine are an important source of food of the cats and dogs. They are gregarious animals which keep in contact at night by grunting, or by smell in the case of the peccaries which possess a musk gland just above the tail. Their eyesight is generally poor, but their hearing and sense of smell are acute, and in their sounders they are dangerous adversaries, fighting courageously with their tusks. In tropical America the two species of peccary—the collared and white-lipped—are one of the mainstays of the jaguar and puma, which await their chance to take a

Most wild pigs are mainly nocturnal. Those that were not, like the warthog, are rapidly becoming active at night in areas where they are severely persecuted by hunters

straggler from the group rather than foolishly attempt a direct confrontation. The rarest pigs are strangely the largest: the giant forest hog of the central African forests which weighs up to 500lb, and the smallest: the pygmy hog of northern India which weighs about 20lb.

CHAPTER 12

The Most Intelligent Nightshift

The order of primates is characterised by the extensive cerebral hemispheres of the brain, reaching their greatest development in man. During the course of evolution the primates could be said to have concentrated on brain development and intelligence. The majority are diurnal, unlike their predecessors, the insectivores. Out of a total of approximately 175 species, about thirty-two are nocturnal. These are to be found only amongst the most primitive primate families. The development of diurnal vision, and thence colour vision, has reached its most advanced stages in the apes and man.

Unlike the diurnal primates, the nocturnal species are strictly arboreal and seldom venture to the ground. They are mainly restricted to the tropical and subtropical regions. The shrinking of the muzzle and forward placement of the eyes, which is clearly shown in the arboreal lorises, pottos and bushbabies, permits the binocular vision so important to tree living and to leaping animals. With the exception of Madagascar, where twelve species of nocturnal primates occur, the rest of the warmer regions of the world are poorly represented. There are seven species throughout the tropical New World, seven in Africa and only six species in the whole of forested tropical Asia, from India across to Indonesia and northwards to the Philippines.

Sight is well developed in all the night-active relatives of

180

the monkeys, and is the most important sense, followed by hearing, which is also highly evolved in most species, particularly the tropical American ones where powerful vocalisations play an important part in their daily lives. These—the howler monkeys and the douroucouli—are the most monkey-like of the nocturnal primates, whereas the tail-less lorises of the Old World bear the least resemblance to monkeys.

LORISES

The lorises' eyes are their most prominent feature, the overlap of their visual field resulting in increased powers of stereoscopic sight. Their pupils contract to a vertical slit in bright light. The slender loris, a solitary, totally nocturnal animal, occurs only in the forested regions of southern India and Ceylon. Less ponderous than its close relative the slow loris, it can move relatively quickly when necessary and the opposable first digits of each limb enables it to grip branches firmly. Slender lorises are stalkers, and creep silently upon insects, which they seize with both hands. They have large, thin, naked ears and their sense of hearing is well developed; they sleep by day in a tree hollow or curled on a branch, with their heads tucked between their legs.

The heavier slow loris has a much larger range, from northern India to Indo-China and Malaysia, thence across to the Indonesian islands and northwards to the Philippines. Their habits are similar to the slender loris, but they are far more ponderous in their movements. Their thumbs and toes are at right angles to the other digits, resulting in good opposability, and well-developed muscles allow their feet and hands to grip so tightly they are difficult to loosen. They seldom free more than two feet or hands at a time from their holds. Slow lorises are mainly insectivorous, but also eat bird's eggs and fledglings, small lizards and amphibians, and ripe fruit. They have excellent nocturnal vision, their wide pupils becoming

In addition to excellent nocturnal vision, the slender loris has well-developed hearing. An arboreal stalker, it creeps silently upon its insect prey.

More ponderous than its slender relative, the slow loris also has excellent night vision and hearing. A special adaptation of its digit muscles gives it a vice-like grip

vertically elliptical in contraction to protect the eye from bright light.

ANGWANTIBOS AND POTTOS

Like the Asiatic rain forest, the jungle of West Africa contains two very similar loris-like arboreal creatures, the tail-less angwantibo and the potto. Both have extreme opposability of the hands and feet, reduced index fingers and a very strong grip. Little is known of the habits of the angwantibos, as they remain high in the trees of the tall deciduous forest and are seldom seen even by the natives. Their feet and hands are adapted for grasping for long periods without changing position, and when sleeping the angwantibo's limbs are apparently disconnected from the nervous system, becoming quite numb and showing no sign of life. Its grip on branches is so tight that even in death it may not be released.

The potto is distinguished from the angwantibo and the Asiatic lorises by its short tail, which is little more than an inch long. Pottos' large toes and thumbs are opposable to the other digits and form excellent grasping organs, enabling them to cling with such strength that they are difficult to remove without being caused bodily harm. The potto has cervical spines which penetrate the skin; with its head bowed, it is said to present these to a potential enemy as a deterrent. The use of these spines as a defence is, however, doubted by many zoologists. Although slow in their general movements, pottos are quick to seize insects and other moving prey.

BUSHBABIES

The galagos or bushbabies are the other extreme to the lorises. They are active, agile and highly strung animals, capable of making prodigious leaps, of over 20ft in the larger species. The first digits of each limb are opposable to the others, although not so widely as in the lorises, and their grip is not so strong. All the galagos are decidedly nocturnal, although Demidoff's, the smallest species, is said by some field zoologists to be mainly diurnal, apparently because of a pair seen to be active in bright sunlight. Their enormous eyes and ears are, however, characteristic of other nocturnal members of the family.

Galagos can rotate their heads through 180°, and their large hairless ears, which have several transverse ridges, can be moved towards the direction of sounds. Their ears can be folded over, singly or together, when they are sleeping. When awakened during the day, they are usually so drowsy and dazed by bright light that they are easily caught by trappers. When disturbed, particularly from their sleep, their main alarm calls are a low growling and chattering. Insects, small lizards and amphibians, bird's eggs and fledglings are all eaten by the galagos, which vary their diet with fruit, berries, shoots and flowers. They sleep in tree hollows or in the crook of a tree, and occasionally in the open, shielded by dense foliage and with their heads curled down between their forelimbs. On the rare occasions when they have been seen on the ground, they bounce along like kangaroos, holding their tails

The angwantibo, a tail-less nocturnal primate of the West African forests, is still something of a mystery and a rarity. Seldom seen in the wild, very few have reached zoological garden or museum collections

stiffly behind them as balancing organs.

The Senegal galago is the most widespread species, occurring in a number of forms throughout the savannah woodlands of central and southern Africa. They usually associate in family groups, sleeping in hollow trees or in large leafy nests often built just a few feet from the ground. The Moholi galago of South Africa is the best-known form, and the one most frequently seen in captivity. Its deep topaz eyes gleam intensely red when caught in a beam of artificial light. Allen's galago is similar, but has a longer muzzle and reddish colouring on the hindlimbs; unlike the Senegal bushbaby, it is a true forest species, occurring in the primeval forests of West Africa.

The thick-tailed galago is the largest species and the most carnivorous of all. It has been known to attack tree-roosting

Sleeping by day in tree hollows or leafy nests, often in family groups, the Senegal bushbaby or galago is active and agile at night. Rotating its head 180° it turns its large hairless ears in the direction of a sound, and makes a prodigious leap to capture an insect

or nesting birds, such as the bush partridge, francolin and even guineafowl, often satisfying its hunger by eating only the brain. This bushbaby sleeps in family groups in tree holes and is not known to build nests.

Demidoff's galago is by far the smallest species, being little larger than a mouse, with short ears, large eyes and a very broad head. Despite its small size, this active little animal is capable of horizontal leaps of 6ft. Its call is said to be a short rising chatter, repeated every ten seconds, and so powerful that it is audible a hundred yards away. They have a variety of other calls, by which they communicate in the great forests of West Africa, an indication that they must have fairly acute hearing. Demidoff's bushbaby is a tree nester, either in nests of its own making or in abandoned squirrels' dreys.

TARSIERS

The tarsiers of Indonesia, insular Malaysia and the Philippines are immediately distinguishable from the galagos by their long naked tails. Their name is derived from the elongated tarsal or ankle region, and extreme opposability of the digits is only possible in the feet. Apart from their rat-like tails, huge eyes are their most noticeable feature, and in profile their convex corneas stand out from their faces. The tarsier's pupils show rapid reactions to changes in illumination. Modifications have resulted in powerful eye muscles which enable them to open their eyes to their widest at night, the great enlargement of the pupil reducing the iris to a narrow band. They tilt their heads sideways to watch moving objects and to bring them into focus, and can rotate their heads through almost a complete circle—practically 180° each way—enabling them to see clearly behind without altering the body position. Tarsiers have large, paper-thin, concave pinnae, which are heavily ridged; they move these constantly, like the bats, to locate the source of a noise. As artificial high-pitched sounds have produced rapid ear movement, it is assumed that they are perceptible to the tarsiers. Their own vocalisations are high-pitched twitterings.

At the ends of their digits the tarsiers have expanded discs, which enable them to cling to branches as they jump like tree frogs through the trees at night. The tail acts as a support when they are resting and is used as a balancing organ when they jump. They cling vertically to branches when asleep.

LEMURS

The lemurs, one of the most primitive families of primates, have more nocturnal members than any other primate group. They are restricted to Madagascar and the neighbouring Comoro Islands, where they have evolved free from interference or competition from the higher forms of animal life. Superstitions surround many of them and these species are avoided by the natives, but others are, or were in their more plentiful days, an important source of protein. The decline of many species is attributed to human predation for the pot among the now limited numbers resulting from forest denudation.

The nocturnal lemurs have a more pointed muzzle and less enlarged eyes than the lorises or tarsiers. As the eyes are located more on the sides of the head, there is less overlapping and therefore a lesser degree of binocular vision. Exceptions are the aye-aye, avahi and dwarf lemurs. Studies have indicated that most lemurs are colour blind, but it is thought that there is a varying degree of colour perception in some species.

The smallest lemurs—the dwarf and mouse lemurs—are purely arboreal and nocturnal, and usually have a season of torpidity. They sleep during the day, and when aestivating in the dry season, in a hollow tree or tree nest of their own making. The mouse lemur vies with the pygmy marmoset for the title of the smallest primate. It is squirrel-like in form, with large eyes and large, thin membranous ears, and occurs in the forests and in dense lakeside reed beds. It is said to have a loud, shrill call.

The dwarf lemurs have large eyes placed forwards and together for binocular vision. Their ears are similar to those of the mouse lemur although smaller, and the face in profile is more flattened. The greater dwarf lemurs and the hairy-eared dwarf lemurs sleep in tree holes by day, and judging from the excessively fat condition of some specimens when

Powerful eye muscles enable the tarsier to open its large eyes their widest at night, and to quickly contract the pupils in bright light

collected are believed to aestivate. The remaining species—the fat-tailed dwarf lemur—stores food in its tail for use during the dry season, but there is no evidence that it aestivates like the others, individuals having been seen throughout the year, scurrying through the trees like squirrels. They are said to sleep in company with other nocturnal lemurs.

The two species of gentle lemurs are completely nocturnal denizens of the high altitude bamboo groves and reed beds, their main source of food being derived from these plants. Their procumbent bottom incisors are used as scrapers on the stems and, as they are completely herbivorous, they spend the larger part of their active hours seeking food. The gentle lemurs have short ears, long stiff vibrissae and medium-sized eyes, and their calls have been described as pig-like grunts. Two other unusual nocturnal lemurs are the weasel lemur and the sportive lemur, both secretive, agile animals seldom seen even at night. The former lives in the forests of the humid eastern slopes of the central mountain range, and the sportive lemur occurs in the western and southern forests.

The avahi is a very rare member of the Indriidae family, the only nocturnal member in fact. In consequence of its habits, its pupils are contractile and become vertical slits in

The ruffed lemur is the most nocturnal of the true lemurs, but still enjoys basking in the sun in the tree tops. A native of Madagascar it is now very rare due to forest destruction

Often called night monkeys, douroucoulis are well adapted for nocturnal survival. The retinas of their large eyes are completely devoid of cones, but are rich in rods which are sensitive to light intensity

daylight, and its large eyes are more forward facing than those of the other members of its family. The avahi is usually seen in an upright position when pursued. It is almost entirely vegetarian in its eating habits, living upon leaves, shoots and flowers.

The most unusual of all the lemurs, whether diurnal or nocturnal, is without doubt the aye-aye, a highly specialised arboreal species adapted for gathering wood-boring insect larvae, which form the bulk of its food. It also eats eggs and a certain amount of vegetable matter, notably bamboo pith. Aye-ayes have large, well-developed eyes, with pupils which contract to a vertical slit in bright light. Their large membranous and naked ears indicate a good sense of hearing, confirmed by their ability to locate insect larvae below the surface. The aye-aye listens for sounds of boring, occasionally tapping with its elongated middle finger, possibly to locate hollow areas. This long finger is then inserted into the tunnel and the grub removed on the hooked claw. When the prey is out of reach, the wood is chewed away with the large incisors. The daylight hours are spent in tree holes or in a thick tree fork.

DOUROUCOULIS

The remaining nocturnal primates are confined to tropical America and all except one are howler monkeys. The odd monkey out in this group is the douroucouli, often confusingly known as the night monkey, a name also given to the galagos. They are slender monkeys, reminiscent of the owls in their quiet movements, and even land silently after prodigious leaps, their long tails curling beneath branches and acting as a brake when they land. Nocturnalism has even greater survival value than normal for some of the more southerly douroucoulis. The water-loss of those inhabitating the dry bush of the Gran Chaco in southern South America, for instance, was found to be insignificant due to their nocturnal habits, and their water needs were provided by dew and vegetarian diet. Douroucoulis' eyes are well adapted for nocturnal vision, having no cones at all in their retinas. Their pupils in contraction are circular, showing the conspicuous irides to advantage.

Douroucoulis are active and noisy at night, their power of voice being out of all proportion to their size. They have several calls, one having been likened to the roar of a large cat, another to the booming roar of the howler monkeys. An inflatable chin sac increases the resonance of the voice. They roost communally in tree holes during the day, but are very inquisitive and can usually be relied upon to peer out when awakened by unusual noises below their tree.

HOWLER MONKEYS

Despite their name, the howler monkeys seldom howl, at least not in the sense that dogs howl. Their calls are usually a low roaring sound, rather like that of a distant waterfall. They call intermittently throughout the day when they are resting and at night as they search the branches of their territory for leaves, shoots, berries and green fruits. Denizens of the tropical rain forest from Mexico southwards to Paraguay, they are also found on Trinidad. Their lower jaw and hyoid bone, or tongue bone, are enlarged, giving them great power of voice. They roar in defence of their territory, to outline their area to neighbouring troops, to direct the movement and behaviour of their clan, and for many other reasons. Their auditory powers are assumed to be well developed too, to enable them to hear the calls of neighbouring troops some miles away. Sight is also well developed, and their sense of smell is considered to be good, on account of their strong body odour, which has been likened to garlic.

Howler monkeys' territorial areas are said to average

approximately 250 acres, and rivalry between neighbouring troops, where territories overlap, results in howling contests instead of fighting. Their calls carry at least four miles when conditions are ideal. Troop size varies from four or five individuals of one family to groups totalling thirty animals. Only six species are recognised, one being rufous red, another yellowish-brown and the remainder dark brown to black.

Bibliography

Bogert, C. M. 'How Reptiles Regulate Their Body Temperature', *Scientific American*, 200 (1959), 105

Bourliere, F. *The Natural History of Mammals* (New York 1954)

Cahalane, V. H. *Mammals of North America* (New York 1947)

Cochrane, D. M. *Living Amphibians of the World* (New York 1961)

Crile, G. *Intelligence, Power and Personality* (New York 1941)

Ditmars, R. L. *Reptiles of the World* (New York 1943)

——and Greenhall, A. M. 'The Vampire Bat', Smith Report for 1936 (Washington, DC). 277–96

Griffin, D. R. *Listening in the Dark* (Yale Univ Press 1958)

Hill, W. C. O. *Primates, Comparative Anatomy and Taxonomy*, vols 1–5 (1953–62)

Kruuk, H. and Turner, M. 'Comparative Notes on Predation by Lion, Leopard, Cheetah and Wild Dogs in the Serengeti', *Mammalia*, 31 (1967), 1–27

Matthews, L. H. and Knight, M. *The Senses of Animals* (1963)

Milne, L. and M. *The Senses of Animals and Men* (New York 1962)

Minton, S. A. Jr and M. R. *Venomous Reptiles* (New York 1969)

Bibliography

Ognev, S. I. *Mammals of Eastern Europe and Northern Asia*, vol 1 (Jerusalem 1962)

Prater, S. H. *The Book of Indian Animals* (Bombay 1965)

Roberts, A. *The Mammals of South Africa* (1954)

Schmidt, K. P. and Inger, R. F. *Living Reptiles of the World* (New York 1957)

Smythe, R. H. *Animal Vision* (Springfield, Ill. 1961)

Thompson, A. L. (ed). *A New Dictionary of Birds* (1964)

Troughton, E. *Furred Animals of Australia* (Sydney 1967)

Walker, E. P. *Mammals of the World*, vol 1–2 (Baltimore 1964)

Wright, A. H. and Wright, A. A. *Handbook of Frogs and Toads* (Ithaca, NY 1967)

Index